Lecture Notes in Mathematics 2025

Editors:
J.-M. Morel, Cachan
B. Teissier, Paris

Subseries:
École d'Été de Probabilités de Saint-Flour

T0074115

For further volumes:
http://www.springer.com/series/304

Saint-Flour Probability Summer School

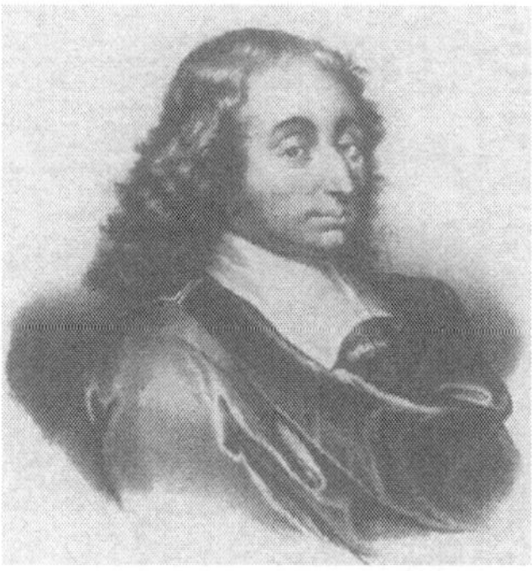

The Saint-Flour volumes are reflections of the courses given at the Saint-Flour Probability Summer School. Founded in 1971, this school is organised every year by the Laboratoire de Mathématiques (CNRS and Université Blaise Pascal, Clermont-Ferrand, France). It is intended for PhD students, teachers and researchers who are interested in probability theory, statistics, and in their applications.

The duration of each school is 13 days (it was 17 days up to 2005), and up to 70 participants can attend it. The aim is to provide, in three high-level courses, a comprehensive study of some fields in probability theory or Statistics. The lecturers are chosen by an international scientific board. The participants themselves also have the opportunity to give short lectures about their research work.

Participants are lodged and work in the same building, a former seminary built in the 18th century in the city of Saint-Flour, at an altitude of 900 m. The pleasant surroundings facilitate scientific discussion and exchange.

The Saint-Flour Probability Summer School is supported by:

- Université Blaise Pascal
- Centre National de la Recherche Scientifique (C.N.R.S.)
- Ministère délégué à l'Enseignement supérieur et à la Recherche

For more information, see back pages of the book and
http://math.univ-bpclermont.fr/stflour/

Jean Picard
Summer School Chairman
Laboratoire de Mathématiques
Université Blaise Pascal
63177 Aubiére Cedex
France

Giambattista Giacomin

Disorder and Critical Phenomena Through Basic Probability Models

École d'Été de Probabilités de Saint-Flour XL – 2010

Springer

Giambattista Giacomin
Université Paris Diderot
Département de Mathématiques
175 rue du Chevaleret
75205 Paris
France
giacomin@univ-paris-diderot.fr

ISBN 978-3-642-21155-3 e-ISBN 978-3-642-21156-0
DOI 10.1007/978-3-642-21156-0
Springer Heidelberg Dordrecht London New York

Lecture Notes in Mathematics ISSN print edition: 0075-8434
ISSN electronic edition: 1617-9692

Library of Congress Control Number: 2011933580

Mathematics Subject Classification (2011): 82B44, 60K35, 60K37, 82B27, 60K05, 82D30

© Springer-Verlag Berlin Heidelberg 2011
This work is subject to copyright. All rights are reserved, whether the whole or part of the material is concerned, specifically the rights of translation, reprinting, reuse of illustrations, recitation, broadcasting, reproduction on microfilm or in any other way, and storage in data banks. Duplication of this publication or parts thereof is permitted only under the provisions of the German Copyright Law of September 9, 1965, in its current version, and permission for use must always be obtained from Springer. Violations are liable to prosecution under the German Copyright Law.
The use of general descriptive names, registered names, trademarks, etc. in this publication does not imply, even in the absence of a specific statement, that such names are exempt from the relevant protective laws and regulations and therefore free for general use.

Cover design: deblik, Berlin

Printed on acid-free paper

Springer is part of Springer Science+Business Media (www.springer.com)

Preface

These notes are a revised version of the ones that I have prepared and used for my course at the 40th Saint-Flour Probability Summer School. I was extremely happy to receive the invitation and giving the course has been a real pleasure. This marks my my third time participating in the school and this preface is the occasion to compliment Jean Picard for his discrete, smooth and efficient way of running it.

This invitation gave me the opportunity to rethink my research activities in the last 5 or 6 years, trying to take a somewhat different standpoint: in the end I realized that I was just trying to go back to the original motivations. In this period in fact I have been working mostly on localization phenomena in certain disordered systems and a class of models – the pinning models – took a leading position. But the question driving my interest was and is: what is the effect of disorder on phase transitions and on critical phenomena? So the result is that, if we look at these lectures from a technical viewpoint, they are about the specific class of statistical mechanics models that I call disordered pinning models. For this class we do have fairly satisfactory answers: essentially all the physical predictions on which there was a general consensus have now been established on firm mathematical grounds and there are now also rigorous result about some controversial physical statements. But, beyond the purely technical aspects, these notes are also an invitation to look beyond pinning models, that is, to more general statistical mechanics models.

It suffices to browse through these pages to realize that "more general statistical mechanics models" essentially reduces to the Ising model (and this is still, definitely, too much for these notes). The Ising model is going to accompany us along the various steps, but (hopefully!) in a way that it is not too invasive: the reader who is only interested in pinning phenomena should be able to follow leaving aside the sections on the Ising model, in which the presentation is rather informal. The choice of keeping disordered Ising models issues at an informal level is also due to the lack of rigorous results, in spite of some absolutely impressive achievements, and these portions of the notes present a number of open problems, which are most probably really challenging.

I would like to stress that since (polymer, interface, Markov process, etc.) pinning models are the means and not the aim, the modeling aspects and the very many variations or closely related classes of models are reduced to remarks or are even neglected entirely unless directly related to the main line of the notes. In this sense these notes do not review, for example, the vast literature on polymer models, not even that on general pinning or localization phenomena.

Moreover, these notes do not include hierarchical models on diamond lattices. Choices had to be made and this was the most painful one for at least two reasons: on one hand, part of the results and of the phenomena that I present have been obtained or have been understood first in the hierarchical set-up and, on the other, these somewhat exotic models definitely have a particular inner beauty.

I am presenting the combined work of several persons, to whom I am deeply grateful and indebted. I want to especially thank my closest collaborators and the people with whom I have discussed the subject of these notes most: Francesco Caravenna, Hubert Lacoin and Fabio Toninelli. Moreover special thanks are due to Bernard Derrida, who shaped my vision of the Harris criterion and who helped me in going through the physics literature (needless to say, I take full responsibility for what I have written on physics issues and I am absolutely aware that Bernard would have put things differently). I certainly cannot forget that my interest in disordered models and in localization phenomena dates back to my valued interaction, which persisted through the years, with Erwin Bolthausen.

Finally, I would like to give a big "thank you" to Lydia, Micol and Raika for their presence, in Saint-Flour, before and after.

Paris
March 2011

Giambattista Giacomin

Contents

1 **Introduction** 1
References 4

2 **Homogeneous Pinning Systems: A Class of Exactly Solved Models** 5
2.1 What Happens if We Reward a Random Walk When it Touches the Origin? 5
2.1.1 The Random Walk Pinning Model 5
2.1.2 Visits to the Origin and the Computation of the Partition Function 7
2.1.3 From Partition Function Estimates to Properties of the System 10
2.2 The General Homogeneous Pinning Model 14
2.3 Phase Transition and Critical Behavior 18
2.4 A First Look at a Crucial Notion: The Correlation Length 19
2.5 Why Do People Look at Pinning Models? A Modeling Intermezzo 21
2.5.1 Polymer Pinning by a Defect 21
2.5.2 Interfaces in Two Dimensions 21
2.5.3 DNA Denaturation: The Poland–Scheraga Model 24
2.6 A Look at the Literature 24
References 26

3 **Introduction to Disordered Pinning Models** 29
3.1 The Disordered Pinning Model 29
3.2 Super-Additivity, Free Energy, and Localization 32
3.2.1 Two Important Remarks 34
3.3 Self-Averaging Property, Effect of Boundary Condition 35
3.3.1 Proof of Proposition 3.2 35
3.3.2 Free and Constrained Models 37

3.4 A Look at the Literature and, Once Again, Correlation Length Issues 38
References 40

4 Irrelevant Disorder Estimates 41
4.1 Disorder and Critical Behavior: What to Expect? 41
4.1.1 First Approach: An Expansion in Powers of β^2 42
4.1.2 Second Approach: A 2-Replica Argument 44
4.2 Disorder is Irrelevant if $\alpha < 1/2$ (and if β is Not Too Large): A Proof 46
4.3 A Look at the Literature 49
References 50

5 Relevant Disorder Estimates: The Smoothing Phenomenon 51
5.1 Smoothing for Gaussian Charges: The Rare Stretch Strategy 51
5.2 More General Charge Distributions 55
5.3 Back to and Beyond Harris Criterion: Disorder and Smoothing 55
5.3.1 Disorder and Phase Transitions 56
5.3.2 Harris' Heuristic Argument 57
5.3.3 Relevance and Irrelevance 58
5.3.4 The Diluted Ising Model 58
5.3.5 Random External Fields 59
5.4 A Further Look at the Literature 60
References 60

6 Critical Point Shift: The Fractional Moment Method 63
6.1 Main Result and Overview 63
6.2 The Basic Fractional Moment Estimates 65
6.3 The $\alpha > 1$ Case 67
6.3.1 A Different Look on Proposition 6.3 67
6.3.2 A First Coarse Graining Procedure: Iterated Fractional Moment Estimates 68
6.3.3 Finite Volume Estimates: The Proof of Theorem 6.1 for $\alpha > 1$ 70
6.4 The $\alpha = 1$ Case 74
6.5 The $\alpha \in (1/2, 1)$ Case 75
6.5.1 Bounds for Correlation Length Size Systems 76
6.5.2 Proof of Theorem 6.1, Case $\alpha \in (1/2, 1)$ 78
6.6 The $\alpha = 1/2$ Case 79
6.6.1 Estimates up to the (Annealed) Correlation Length: Gaussian Case 79
6.6.2 Beyond the Correlation Length: The Proof of Theorem 6.1 ($\alpha = 1/2$) 83
6.7 A Look at the Literature 87
References 88

7 The Coarse Graining Procedure 91
7.1 Coarse Graining Estimates 91
References 99

8 Path Properties 101
8.1 Overview 101
8.2 A Quick Look at Concentration Inequalities 102
8.3 The Localized Regime 104
8.3.1 A Basic Observation (and its Consequences) 104
8.3.2 On $\mu(\beta,h)$ and $\mathrm{F}(\beta,h)$ 106
8.4 The Delocalized Regime 108
8.5 Path Behavior: Overview of What is Known and What is Not 109
8.5.1 On the Localized (and Critical) Regime 110
8.5.2 On the Delocalized Regime 111
References 111

A Discrete Renewal Theory: Basic (and a Few Less Basic) Facts and Estimates 113
A.1 A Crash Course on Renewal Theory 113
A.1.1 Renewal and Markov Chains 113
A.1.2 The Renewal Theorem 114
A.1.3 Beyond the Renewal Theorem 115
A.1.4 Convergence of Renewal and Point Processes 116
A.2 Some Pinning Oriented Renewal Issues 117
A.2.1 On Boundary Effects 117
A.2.2 Two Scaling Results for Renewal Processes 118
A.2.3 On the Derivatives of the Free Energy Near Criticality 122
References 125

Index 127

List of participants 129

Frequently Used Notations

$\mathbb{N}$	$\{1,2,3,\ldots\}$
$a \wedge b$, $a \vee b$	$\min(a,b)$, $\max(a,b)$
$\|E\|$, $\mathscr{P}(E)$, $\mathbf{1}_E$	Cardinality, set of all subsets, indicator function of E
$\{a_n\}_n, \{b_n\}_n, \ldots$	Sequences of real numbers
$a_n \sim b_n$	$\lim_n a_n/b_n = 1$
$a_n \approx b_n$	Used when one does not want, or cannot, be precise
$\tau = \{\tau_j\}_{j=0,1,\ldots}$	Renewal sequence, often seen as subset of $\mathbb{N} \cup \{0\}$
$S = \{S_j\}_{j=0,1,\ldots}$	Random walk
$\mathbf{P}$	Law of τ or law of S, according to the context
$K(n)$	$\mathbf{P}(\tau_1 = n)$, $n = 1,2,\ldots,\infty$ (Chap. 2, Sect. 2.2)
$\overline{K}(n)$	$\sum_{j>n} K(n) \leq 1$ ($n = 0,1,\ldots$, sum does not include ∞)
$\widetilde{K}_h(n)$	(2.10) and Chap. 2, Sect. 2.2
$\widetilde{\tau}^{(h)}$	Renewal with inter-arrival law $\widetilde{K}_h(n)$
θ	Left shift operator: $(\theta a)_n = a_{n+1}$
h	Pinning potential
δ_n	$\mathbf{1}_{n\in\tau}$ (abuse of notation for $\mathbf{1}_\tau(n)$, here $\tau \subset \mathbb{N} \cup \{0\}$)
$Z_{N,h}$, $Z^{\mathbf{f}}_{N,h}$	Partition function of homogeneous system (constrained, free)
$\mathbf{P}_{N,h}$, $\mathbf{P}^{\mathbf{f}}_{N,h}$	Probability law of homogeneous system (constrained, free)
$\mathrm{F}(h)$	Free energy of model with homogeneous pinning potential h
κ	correlation length
ω, $\mathbb{P}$	disorder or charge sequence, law of ω: Definition 3.1
$\mathrm{M}(\beta)$	$\mathbb{E}\exp(\beta\omega_1)$: Definition 3.1
β	disorder strength parameter
$Z_{N,\omega} = Z_{N,\omega,\beta,h}$	Partition function of disordered system (constrained)
$Z^{\mathbf{f}}_{N,\omega} = Z^{\mathbf{f}}_{N,\omega,\beta,h}$	Partition function of disordered system (free)
$\mathbf{P}_{N,\omega} = \mathbf{P}_{N,\omega,\beta,h}$	Disordered system probability (constrained)
$\mathbf{P}^{\mathbf{f}}_{N,\omega} = \mathbf{P}^{\mathbf{f}}_{N,\omega,\beta,h}$	Disordered system probability (free)
$\mathrm{F}(\beta,h)$	Free energy of the disordered pinning model ($\mathrm{F}(0,h) = \mathrm{F}(h)$)
IID	Independent and Identically Distributed

Chapter 1
Introduction

Disorder enters modeling in a very natural way: interacting "units" (spins, particles, circuits, cells, individuals,...) are not identical, media (solvents, lattices, environments,...) are not homogeneous or regular, and so on. In many instances it is of course very reasonable to assume that heterogeneities, irregularities, impurities,... can be neglected, and even more for toy models. But it is at least as reasonable to wonder about the stability of the results one obtains for homogeneous systems if disorder is introduced. This concern is omnipresent in the scientific literature and several instances in which small impurities have a drastic effect have been exposed. Four examples for all:

1. Anderson localization in one-dimensional systems, where an arbitrary amount of disorder transforms a conductor into an insulator, see [12] and references therein, and one can find a vast physical literature on the analogous phenomenon in dimension two.
2. Sinai's random walk in random environment, where strongly sub-diffusive behavior sets in for arbitrarily weak randomness in the environment ([13] and references therein).
3. Directed polymers in random environments in dimension one and two, where once again arbitrarily small amounts of disorder lead to drastic phenomena, like, at least in dimension one, super-diffusivity and non Gaussian fluctuations (see [1,3,5] and references therein also for the link to other classes of models in the same "universality class").
4. The phase transition in the Ising model can be strongly affected by arbitrarily weak disorder: it may even disappear.

These notes are very close in spirit to all these four remarkable examples – for example, they share with the first three the central keyword "localization" – but the reader will hardly find traces of examples one to three in the sequel. The situation is radically different for the fourth example: these notes contain several discussions and references on disordered Ising models (that is why we have put no reference here!), even if the Ising model is not their main subject. Rather, pinning models are: let us explain why (and what pinning is).

G. Giacomin, *Disorder and Critical Phenomena Through Basic Probability Models*, Lecture Notes in Mathematics 2025, DOI 10.1007/978-3-642-21156-0_1,

© Springer-Verlag Berlin Heidelberg 2011

The main purpose of these notes is to explore the effect of disorder in the framework of equilibrium statistical mechanics of lattice models, that is for Gibbs measures with random interactions or random external fields. These issues have been first developed for the reference statistical mechanics model – the Ising model – and soon after Onsager's celebrated exact solution of the two dimensional case researchers started wondering about the stability of Onsager's result when impurities are introduced. We want to give an overview of the remarkable ideas developed in this context, but trying to follow in detail the Ising model literature would be rather pretentious (and beyond the possibilities of the author): we invite the reader to have a look at the references at the end of Chap. 5 to get a first idea of the body of literature available and we refer to [2] for a comprehensive recent reference on disordered models. The reason why a full account of the Ising literature would be a daunting enterprise is not only due to the amount and depth of the results available, but also to the fact that a consistent part of the physical predictions for the moment are not on firm mathematical grounds. This is the case in particular for the work of Harris [10] on the diluted Ising model and of the various works on disorder relevance/irrelevance that followed. In fact, Harris' idea – that yields the so called Harris criterion – is that one should be able to predict whether a small amount of disorder – "impurities" in Harris' terminology – changes or not the critical behavior of a system, that is the behavior of a system near the phase transition, by simply looking at the critical behavior of the "pure" system. This is very much in the spirit of perturbation theory, but it is a delicate issue because one is dealing with infinite systems and even if locally one adds a small amount of impurities, one always does it in a statistically translation invariant fashion, so in the end the amount of disorder is infinite anyway.

However, if for Ising the full mathematical picture is still escaping, there is a class of models – the pinning models – in which Harris' ideas have been fully understood. Moreover, homogeneous pinning models display phase transitions of "all orders": in fact, by playing on a parameter of the model, one can observe any type of critical behavior of the order parameter at the transition, while for the Ising model one has a discrete spectrum of possible behaviors (the parameter is the dimension), not to speak of the fact that the precise Ising critical behavior is still an open problem in some dimensions. So the basic scheme of the lectures can be summed up to

- Developing in detail the analysis of pinning models
- Discussing, in a less technical fashion, what happens, or what is expected to happen, for the Ising model

But what are pinning models? Just think of an arbitrary, say discrete time, Markov process that visits with positive probability a state (call it 0): for example a random walk on $\mathbb{Z}^d$, with symmetric or asymmetric IID increments, that jumps to nearest neighbor sites. It is well know that an asymmetric walk is transient, so any site, 0 in particular, is visited only a finite number of time if any: moreover, it has a definite drift in a direction. A symmetric walk instead is transient or recurrent according to whether $d \geq 3$ or $d \leq 2$, and in any case there is no drift direction: the walk diffuses. What happens if we reward visits to 0? More precisely, what happens if we modify the law of the walk (up to time N) by weighing the probability of each configuration with the exponential of a constant h times the time spent in 0 up to time N? And we are interested in the limit $N \to \infty$. The answer in general is that there is a delocalization/localization transition, that is for h above a certain h_c the walk localizes at 0 (it becomes positive recurrent), while below h_c the walk visits 0 no more than a finite number of times (transient behavior). The transition can be characterized in terms of the contact fraction, that is the number of returns to 0 in a long stretch of time, divided by the length of the stretch: zero contact density is delocalization and positive contact density is localization. In this case the critical behavior is the way the contact fraction approaches 0 as $h \searrow h_c$ (when it does, because we will see that in some cases it jumps to zero: this is the case of the so-called "first order transitions").

The pinning model we have informally introduced is homogeneous: the reward (or penalty) h is constant ($h > 0$ is a reward, $h < 0$ is a penalty). In the disordered version h is not constant, in fact h is replaced by $h + \beta\omega_n$ with $\beta > 0$ and $\omega := \{\omega_n\}_{n\in\mathbb{N}}$ is a realization of a sequence of random variables (for example, independent and identically distributed: in these notes we will only consider this case, and we will consider ω_1 centered and of variance one). We insist on the fact that we choose a typical realization of ω and we keep it fixed: a walk touching 0 at time n will receive a reward (or penalty) $h + \beta\omega_n$. And here is the main question of these notes: how different are the $\beta = 0$ and the $\beta > 0$ case? We will see that for pinning models the issue is not so much the one of persistence of the phase transition, because the answer is going to be positive in all cases, but the one of whether the disorder has an effect on the critical behavior or not.

Here is an overview of what will follow:

- In Chap. 2 we introduce and solve a general class of homogeneous pinning models. The emphasis is on the renewal process viewpoint because we are mostly interested in (in these notes!) when the process comes back to 0 and not so much on what it does outside of 0. And the return times to 0 form a random sequence that is a renewal sequence. In this chapter the reader will find also a quick review of the modeling aspects: this is definitely not doing justice to the importance of pinning phenomena and pinning modeling, but this is really not the purpose of the notes, so we just refer to [6] and to the monographs [8, 11].
- In Chap. 3 we introduce the disordered pinning models and a number of basic techniques. Notably we introduce the notion of disordered (quenched) free energy.
- In Chap. 4 there is the first contact with the Harris criterion and with the seminal contributions [4, 7], of which we will present the basic ideas. We then give a mathematical proof of disorder irrelevance in agreement with the Harris criterion prediction.
- In Chap. 5 we show that disorder is relevant when predicted by the Harris criterion. We do this by establishing a bound on the quenched free energy that shows that it always possesses a certain minimal amount of smoothness: it is at

least C^1 with Lipschitz derivative. We then present an overview of what is known or is expected to hold for the Ising model.
- In Chap. 6 we study the shift of the critical point. This includes the analysis of the case of marginal disorder, i.e. neither relevant nor irrelevant, a debated issue in the physical literature, solved in [9].
- In Chap. 7 we prove the coarse graining estimates used in Chap. 6.
- In Chap. 8 we talk about path properties and show that they are tightly linked to the free energy.
- The appendix is about discrete renewal processes and it is split into two parts: in the first part we review some basic tools and results of the theory and the second part is about some issues that are more specific to pinning models.

References

1. M. Balázs, J. Quastel, T. Seppäläinen, Fluctuation exponent of the KPZ/stochastic Burgers equation. J. Am. Math. Soc. (2011), published online
2. A. Bovier, *Statistical Mechanics of Disordered Systems. A Mathematical Perspective*. Cambridge Series in Statistical and Probabilistic Mathematics (Cambridge University Press, Cambridge, 2006)
3. F. Comets, N. Yoshida, T. Shiga, Probabilistic analysis of directed polymers in random environment: a review. Adv. Stud. Pure Math. **39**, 115–142 (2004)
4. B. Derrida, V. Hakim, J. Vannimenus, Effect of disorder on two-dimensional wetting. J. Stat. Phys. **66**, 1189–1213 (1992)
5. P. Ferrari, H. Spohn, *Random growth models*, in The Oxford Handbook of Random Matrix Theory, ed. by G. Akemann, J. Baik, P. Di Francesco (2011)
6. M.E. Fisher, Walks, walls, wetting, and melting. J. Stat. Phys. **34**, 667–729 (1984)
7. G. Forgacs, J.M. Luck, Th. M. Nieuwenhuizen, H. Orland, Wetting of a disordered substrate: exact critical behavior in two dimensions. Phys. Rev. Lett. **57**, 2184–2187 (1986)
8. G. Giacomin, *Random Polymer Models* (Imperial College Press, London, 2007)
9. G. Giacomin, H. Lacoin, F.L. Toninelli, Marginal relevance of disorder for pinning models. Commun. Pure Appl. Math. **63**, 233–265 (2010)
10. A.B. Harris, Effect of random defects on the critical behaviour of Ising models. J. Phys. C **7**, 1671–1692 (1974)
11. F. den Hollander, *Random polymers*, in Lectures from the 37th Probability Summer School Held in Saint-Flour, 2007, Lecture Notes in Mathematics, vol. 1974 (Springer, Berlin, 2009)
12. L.A. Pastur, *Spectral properties of random selfadjoint operators and matrices (a survey)*, in Proceedings of the St. Petersburg Mathematical Society, vol. IV, pp. 153–195, Am. Math. Soc. Transl. Ser. 2, vol. 188 (Am. Math. Soc., Providence, RI, 1999)
13. O. Zeitouni, *Random walks in random environment*, in Part II in Lectures from the 31st Summer School on Probability Theory Held in Saint-Flour, 8–25 July 2001, ed. by J. Picard, Lecture Notes in Mathematics, vol. 1837 (2004), pp. 189–312

Chapter 2
Homogeneous Pinning Systems: A Class of Exactly Solved Models

Abstract We introduce a class of statistical mechanics non-disordered models – the homogeneous pinning models – starting with the particular case of random walk pinning. We solve the model in the sense that we compute the precise asymptotic behavior of the partition function of the model. In particular, we obtain a formula for the free energy and show that the model exhibits a phase transition, in fact a localization/delocalization transition. We focus in particular on the critical behavior, that is on the behavior of the system close to the phase transition. The approach is then generalized to a general class of Markov chain pinning, which is more naturally introduced in terms of (discrete) renewal processes. We complete the chapter by introducing the crucial notion of correlation length and by giving an overview of the applications of pinning models. Ising models are presented at this stage because pinning systems appear naturally as limits of two dimensional Ising models with suitably chosen interaction potentials. In spite of the fact that these lecture notes may be read focusing exclusively on pinning, the physical literature on disordered systems and Ising models cannot be easily disentangled. So a full appreciation of some physical arguments/discussions in these notes does require being acquainted with Ising models.

2.1 What Happens if We Reward a Random Walk When it Touches the Origin?

2.1.1 *The Random Walk Pinning Model*

We start rather abruptly by making more precise the question in the title and by answering it. So let us give ourselves a random walk $S = \{S_0, S_1, \ldots\}$ with $S_0 = 0$ and such that the increment variables $\{S_n - S_{n-1}\}_{n\in\mathbb{N}}$, that form an IID sequence, take values -1, 0 and $+1$. More precisely we consider a symmetric walk and set

G. Giacomin, *Disorder and Critical Phenomena Through Basic Probability Models*, Lecture Notes in Mathematics 2025, DOI 10.1007/978-3-642-21156-0_2,

© Springer-Verlag Berlin Heidelberg 2011

$\mathbf{P}(S_1 = +1) = \mathbf{P}(S_1 = -1) =: p/2$ and $\mathbf{P}(S_1 = 0) = q$. Of course $p+q=1$: we exclude the trivial case $q=1$ and the simple random walk $q=0$ for its somewhat unpleasant periodic character. For every $N \in \mathbb{N}$ we introduce the *local time* $L_N(S) = \sum_{n=1}^{N} \mathbf{1}_{S_n=0}$ and the probability measure $\mathbf{P}_{N,h}$ ($h \in \mathbb{R}$) such that

$$\frac{\mathrm{d}\mathbf{P}_{N,h}}{\mathrm{d}\mathbf{P}}(S) = \frac{1}{Z_{N,h}} \exp\left(hL_N(S)\right) \mathbf{1}_{S_N=0}, \tag{2.1}$$

where $Z_{N,h}$ is typically called *partition function* and it is just the normalization that makes $\mathbf{P}_{N,h}$ a probability. Of course

$$Z_{N,h} = \mathbf{E}\left[\exp\left(hL_N(S)\right); S_N = 0\right]. \tag{2.2}$$

A word about an abuse of notation that, in different forms, will be ubiquitous in these notes: in (2.1) S is a trajectory of the random walk, rather than the sequence of random variables. Note moreover that we have introduced $\mathbf{P}_{N,h}$ as a measure on the full trajectory and not just for the part of the trajectory that we have really modified. This has plenty of almost irrelevant advantages that, added up, largely overcome (in the eyes of the author, of course) the disadvantage of a rather abstract formulation in terms of the relative density of measures. Note in fact that for every $s_1, s_2, \ldots, s_N$

$$\begin{aligned} &\mathbf{P}_{N,h}\left(S_1 = s_1, S_2 = s_2, \ldots, S_N = s_N\right) \\ &\quad = \frac{\mathbf{1}_{s_N=0}}{Z_{N,h}} \exp\left(h \sum_{n=1}^{N} \mathbf{1}_{s_n=0}\right) \mathbf{P}\left(S_1 = s_1, S_2 = s_2, \ldots, S_N = s_N\right), \end{aligned} \tag{2.3}$$

and we could have used the right-hand side of this expression to define the process, at the expense of having a family of processes living on different spaces and of a less compact notation. A last observation on notation is that $\mathbf{1}_{S_N=0}$ is used in place of the more precise, but less *expressive*, $\mathbf{1}_{\{0\}}(S_N)$.

Remark 2.1. Why constraining to $S_N = 0$? $S_N = 0$ is a boundary condition and a priori it is more natural to introduce the *free*, or *unconstrained*, model

$$\frac{\mathrm{d}\mathbf{P}^{\mathrm{f}}_{N,h}}{\mathrm{d}\mathbf{P}}(S) = \frac{1}{Z^{\mathrm{f}}_{N,h}} \exp\left(hL_N(S)\right), \tag{2.4}$$

but the *constrained* model often turns out to be more manageable. We anticipate that, even if for most of the main results there will be little or no difference between the two models, the interplay between them plays a role in several proofs.

2.1.2 Visits to the Origin and the Computation of the Partition Function

The epochs $\tau = \{\tau_0, \tau_1, \ldots\}$ of successive visits to the origin

$$\tau_0 := 0 \text{ and } \tau_n \overset{n\in\mathbb{N}}{=} \{j > \tau_{n-1} : S_j = 0\}, \qquad (2.5)$$

is a natural random walk (another one! With positive increments this time) associated to the problem, in the sense that $\{\tau_n - \tau_{n-1}\}_{n\in\mathbb{N}}$ is an IID sequence: this is a direct consequence of the strong Markov property and of the recurrent character of S that guarantees that $\mathbf{P}(\tau_j < \infty \text{ for every } j) = 1$. In a more customary terminology, τ is a *renewal process* with *inter-arrival law* $K(n) := \mathbf{P}(\tau_1 = n)$. Two basic facts on this inter-arrival law are

$$\sum_{n=1}^{\infty} K(n) = 1 \text{ and } \lim_{n\to\infty} n^{3/2}K(n) =: c_K > 0, \qquad (2.6)$$

where the first fact is just a restatement of $\mathbf{P}(\tau_1 < \infty) = 1$, but the second requires a bit more work (see e.g. [22, Appendix A.6] where the value of c_K is computed: $\sqrt{p/2\pi}$).

Remark 2.2. We will soon encounter other renewal processes (i.e. random walks with positive increments: Appendix A offers an introduction to these processes). So we introduce some (more or less) standard terminology: a renewal τ with inter-arrival law $K(\cdot)$ will be called $K(\cdot)$-renewal. If $\sum_n K(n) = 1$ then a.s. $|\{j : \tau_j < \infty\}| = \infty$ and the renewal is said *persistent*. It is *positive persistent* if also $\mathbf{E}[\tau_1] = \sum_n nK(n) < \infty$. If instead $\sum_n K(n) < 1$, $K(\cdot)$ can be extended to a probability distribution by setting $K(\infty) := 1 - \sum_{n\in\mathbb{N}} K(n)$ and each realization of τ, still defined as the sequence of partial sums of the IID sequence of variables with distribution $K(\cdot)$ on $\mathbb{N}\cup\{\infty\}$, contains only a finite number of finite numbers (points, epochs,....). In this case we say that the renewal is *terminating*: after a finite number of bounded jumps, the process jumps to infinity and stays there (in a sense, it leaves the space once for all). In general, it is very practical to look at τ as a subset of $\mathbb{N}$, rather than a sequence (in the terminating case we neglect the repeated ∞ and a typical realization of τ is therefore just a finite subset of $\mathbb{N}$, while in the persistent case it is an infinite subset). This convention leads to particularly compact notations: for example $n \in \tau$ means that there exists $j \in \mathbb{N}\cup\{0\}$ such that $\tau_j = n$. It is customary to call $n \mapsto \mathbf{P}(n \in \tau)$ *renewal function*.

By repeated use of the total probability formula we obtain

$$\begin{aligned} Z_{N,h} &= \sum_{n=1}^{N} \mathbf{E}\left[\exp(hL_N(S));\, S_N = 0, L_N(S) = n\right] \\ &= \sum_{n=1}^{N} \exp(hn)\,\mathbf{P}(S_N = 0, L_N(S) = n) \end{aligned}$$

$$= \sum_{n=1}^{N} \exp(hn) \sum_{\ell \in \mathbb{N}^n: |\ell|=N} \mathbf{P}(\tau_1 = \ell_1, \tau_2 - \tau_1 = \ell_2, \ldots, \tau_n - \tau_{n-1} = \ell_n)$$

$$= \sum_{n=1}^{N} \exp(hn) \sum_{\ell \in \mathbb{N}^n: |\ell|=N} \prod_{j=1}^{n} K(\ell_j), \tag{2.7}$$

where $|\ell| = \sum_{i=1}^{n} \ell_i$. The net outcome is:

$$Z_{N,h} = \sum_{n=1}^{N} \sum_{\ell \in \mathbb{N}^n: |\ell|=N} \prod_{j=1}^{n} \exp(h) K(\ell_j). \tag{2.8}$$

Of course $Z_{N,0} = \mathbf{P}(N \in \tau)$, that is the partition function is just the renewal function of τ (see Appendix A), and the right-hand side of (2.8), still for $h = 0$, is a more explicit version of such a function: the point is to apply this observation also when $h \neq 0$. The obstacle is of course that $\exp(h)K(\cdot)$ is no longer a probability distribution if $h \neq 0$: this is not really a serious problem if $h < 0$ since we have seen that it suffices to work on $\mathbb{N} \cup \{\infty\}$, but for $h > 0$ we have to do something different. The idea is to introduce the function $\mathrm{F} : \mathbb{R} \to [0, \infty)$ defined by

$$\sum_{n \in \mathbb{N}} \exp(-n\mathrm{F}(h) + h) K(n) = 1, \tag{2.9}$$

when such a solution exists, that is for $h \geq 0$ (the solution is of course unique by the monotonicity of $x \mapsto \sum_n \exp(-xn)K(n)$). When we cannot solve such a problem, that is for $h < 0$, we set $\mathrm{F}(h) = 0$. Now, for every h we set for $n \in \mathbb{N}$

$$\widetilde{K}_h(n) := \exp(-\mathrm{F}(h)n + h)\, K(n), \tag{2.10}$$

and, adding $\{\infty\}$ if needed, $\widetilde{K}_h(\cdot)$ is a probability distribution.

Remark 2.3. The function $\mathrm{F}(\cdot)$ is called *free energy* and it plays a central role in these notes. A number of properties of $\mathrm{F}(\cdot)$ can be obtained with moderate effort. First of all $\mathrm{F}(\cdot)$ is real analytic except at the origin. The analyticity on the positive semi-axis follows by the Implicit Function Theorem (e.g. [13, Chap. 3, Proposition 2.20]), since $z \mapsto \sum_n K(n) \exp(-zn)$ is analytic on $\{z \in \mathbb{C} : \Re(z) > 0\}$ and its derivative does not vanish on $(0, \infty)$. One verifies directly also that $F(\cdot)$ is convex and that it is increasing on the positive semi-axis: by taking a derivative of the expression in (2.9) and by using the notation $\widetilde{\tau}^{(h)}$ for the $\widetilde{K}_h(\cdot)$-renewal, we see that for $h > 0$

$$\mathrm{F}'(h) = \frac{1}{\sum_n n \widetilde{K}_h(n)} = \frac{1}{\mathbf{E}\widetilde{\tau}_1^{(h)}} > 0, \tag{2.11}$$

and that $\mathrm{F}''(h) = \mathrm{F}'(h)^3 \mathrm{var}(\widetilde{\tau}_1^{(h)}) > 0$.

We can therefore write

$$Z_{N,h} = \exp(\mathrm{F}(h)N) \sum_{n=1}^{N} \sum_{\ell \in \mathbb{N}^n: |\ell|=N} \prod_{j=1}^{n} \widetilde{K}_h(\ell_j), \tag{2.12}$$

a formula that can be made much more compact by using the $\widetilde{K}_h(\cdot)$-renewal $\widetilde{\tau}^{(h)}$:

$$Z_{N,h} = \exp(\mathrm{F}(h)N)\,\mathbf{P}\left(N \in \widetilde{\tau}^{(h)}\right), \tag{2.13}$$

and from such a formula one extracts.

Proposition 2.4. *For the partition functions $Z_{N,h}$ and $Z^{\mathrm{f}}_{N,h}$ defined in* (2.2) *and* (2.4) *we have*

$$\lim_{N\to\infty} \frac{1}{N} \log Z_{N,h} = \lim_{N\to\infty} \frac{1}{N} \log Z^{\mathrm{f}}_{N,h} = \mathrm{F}(h). \tag{2.14}$$

Moreover

$$Z_{N,h} \overset{N\to\infty}{\sim} c_{h,K(\cdot)} \exp(\mathrm{F}(h)N) \times \begin{cases} N^0 & \text{if } h>0, \\ N^{-1/2} & \text{if } h=0, \\ N^{-3/2} & \text{if } h<0, \end{cases} \tag{2.15}$$

with

$$c_{h,K(\cdot)} := \begin{cases} \frac{1}{\sum_n n\widetilde{K}_h(n)} & \text{if } h>0, \\ \frac{1}{\sqrt{2\pi p}} & \text{if } h=0, \\ \frac{c_K \exp(h)}{(1-\exp(h))^2} & \text{if } h<0. \end{cases} \tag{2.16}$$

Proof. The proof of (2.14) in the constrained case is just a matter of showing that $\log \mathbf{P}(N \in \widetilde{\tau}^{(h)}) = o(N)$. But this is obvious since for $h<0$ we have $\exp(h)K(n) \le \mathbf{P}(N \in \widetilde{\tau}^{(h)}) \le 1$ and if $h>0$ (see Fig. 2.1) by the Renewal Theorem (Theorem A.1)

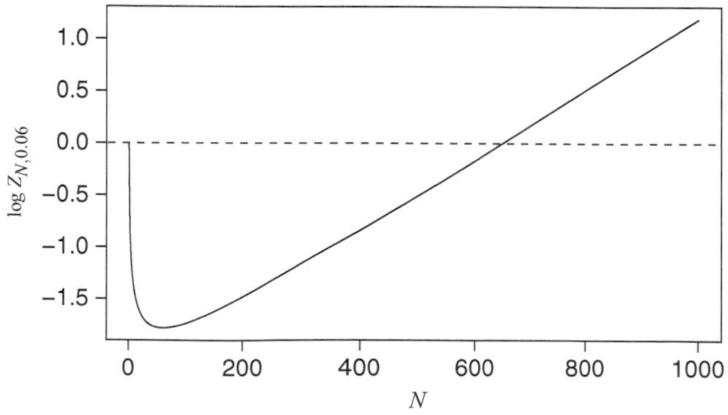

Fig. 2.1 The plot of the logarithm of the partition function for a homogeneous random walk based model with $p = 1/2$ and $q = 1/2$. We have $\mathrm{F}(0.06) \approx 3.4 \times 10^{-3}$

$\mathbf{P}(N \in \widetilde{\tau}^{(h)})$ tends, as $N \to \infty$, to the positive constant $1/\mathbf{E}\widetilde{\tau}_1^{(h)}$: note that this establishes (2.15) for $h > 0$. The sharp estimate (2.15) for $h = 0$ is just the Local Central Limit Theorem for S (which can be established via Stirling's approximation of the factorial), while the case $h < 0$ requires a more delicate analysis, that is however a rather standard result in renewal theory that can be summed up by saying that the leading asymptotic behavior of the renewal function of a terminating renewal differs from the leading asymptotic behavior of the inter-arrival distribution only by a multiplicative constant (see Theorem A.2).

We are left with proving (2.14) in the free case, but this is a direct consequence of the constrained result and of the formula

$$Z^{\mathtt{f}}_{N,h} = \sum_{n=0}^{N} Z^{\mathtt{f}}_{N,h}(\tau \cap [n,N] = \{n\}) = \sum_{n=0}^{N} Z_{n,h}\overline{K}(N-n)\,, \tag{2.17}$$

where we have introduced the notation $Z^{\mathtt{f}}_{N,h}(A)$ (A an event) for $\mathbf{E}[\exp(hL_N(S));A]$ and $\overline{K}(n) = \sum_{j>n} K(j)$ $(n = 0, 1, \ldots)$. From (2.17), and (2.15), one can also establish without much effort the analog of (2.15) for $Z^{\mathtt{f}}_{N,h}$, but this is left to the motivated readers. □

2.1.3 *From Partition Function Estimates to Properties of the System*

Proposition 2.4 contains very detailed information on the system: let us spell it out. First of all we have seen in Remark 2.3 that $\mathrm{F}(\cdot)$ is real analytic except at the origin: convexity assures that at least for $h \neq 0$

$$\mathrm{F}'(h) = \lim_{N\to\infty} \frac{1}{N}\frac{\mathrm{d}}{\mathrm{d}h} \log Z_{N,h} = \lim_{N\to\infty} \frac{1}{N} \mathbf{E}_{N,h}\left[L_N(S)\right]. \tag{2.18}$$

Monotonicity and convexity properties can also be inferred directly form the fact that $h \mapsto \log Z_{N,h}$ is increasing and convex (just take derivatives). Here we are interested in the fact that formula (2.18) is already showing that passing from $h < 0$ to $h > 0$ something very drastic is happening in the system: $\mathrm{F}'(h)$ is actually the density of visits to the origin by the random walk path (the *contact density*) and it passes from zero to a positive value (see Fig. 2.3, upper-right inset). This is clearly a transition from what we may call a delocalized to a localized behavior. The transition actually happens in a continuous way – there is no jump in the contact density when h changes sign – and $\mathrm{F}(\cdot)$ is C^1 in zero, even if it is not C^2: this requires an argument that we develop now. By Riemann sum approximation we see that

$$1-\sum_n K(n)\exp(-xn) = \sum_n K(n)(1-\exp(-xn))$$

$$\stackrel{x\searrow 0}{\sim} c_K x^{1/2}\int_0^\infty \frac{1-\exp(-t)}{t^{3/2}}\mathrm{d}t = 2\sqrt{\pi} c_K x^{1/2}, \quad (2.19)$$

and since we already know that $\lim_{h\searrow 0}\mathrm{F}(h)=0$ we can apply this formula to (2.9) obtaining $2\sqrt{\pi}c_K\mathrm{F}(h)^{1/2}\sim h$, that is

$$\mathrm{F}(h) \stackrel{h\searrow 0}{\sim} \frac{1}{4\pi c_K^2}h^2. \quad (2.20)$$

Such an estimate is directly telling us that $\mathrm{F}(\cdot)$ is not C^2 at the origin and, together with convexity, is telling us also that $\mathrm{F}(\cdot)$ is C^1. In a standard statistical mechanics terminology this means that the system undergoes a *second order phase transition*, in the sense that the non-analiticity of the free energy comes from a singularity (in this case a jump discontinuity) in the second derivative of the free energy.

This description of the system in terms of contact density is only partially satisfactory, for example because we already know that the unperturbed random walk S ($h=0$) has zero contact density, but we know much more, namely that the number of contacts in a stretch N is of order $\sqrt{N}$ (a much sharper information). Can we get such a precise estimate also for the $h\neq 0$ case? The answer is yes and it is summed up in the next statement.

Proposition 2.5. *If $h<0$ then for every $n\in\mathbb{N}$*

$$\lim_{N\to\infty}\mathbf{P}_{N,h}(L_N(S)=n) = (1-\exp(h))^2\, n\exp(h(n-1)), \quad (2.21)$$

(note that the right-hand side is the discrete density of $X+Y+1$, with X and Y independent geometric variables of parameter $\exp(h)$*) while if $h>0$ we have that for every $\varepsilon>0$*

$$\lim_{N\to\infty}\mathbf{P}_{N,h}\left(\left|\frac{L_N(S)}{N}-\mathrm{F}'(h)\right|\geq\varepsilon\right)=0. \quad (2.22)$$

Moreover for every h, every n, every $t_\star\in\mathbb{N}$ and every $t\in\mathbb{N}^n$ such that $0<t_1<t_2<\ldots<t_n\leq t_\star$ we have

$$\lim_{N\to\infty}\mathbf{P}_{N,h}\left(\tau\cap(0,t_\star]=\{t_1,\ldots,t_n\}\right) = \mathbf{P}\left(\widetilde{\tau}^{(h)}\cap(0,t_\star]=\{t_1,\ldots,t_n\}\right), \quad (2.23)$$

that is the sequence $\{\mathbf{P}_{N,h}\tau^{-1}\}_N$ of measures on $\mathscr{P}_0$ (cf. Sect. A.1.4 of Appendix A) converges weakly to $\mathbf{P}(\widetilde{\tau}^{(h)})^{-1}$.

Note that this statement includes global estimates, that is (2.21) and (2.22), and a local one, that is (2.23). Note that (2.22) holds also for $h\leq 0$, but of course for $h\leq 0$ one has much sharper estimates (like (2.21) for $h<0$!). These estimates are just instances of what one can obtain once the sharp asymptotic behavior of the partition function is known (see Fig. 2.2).

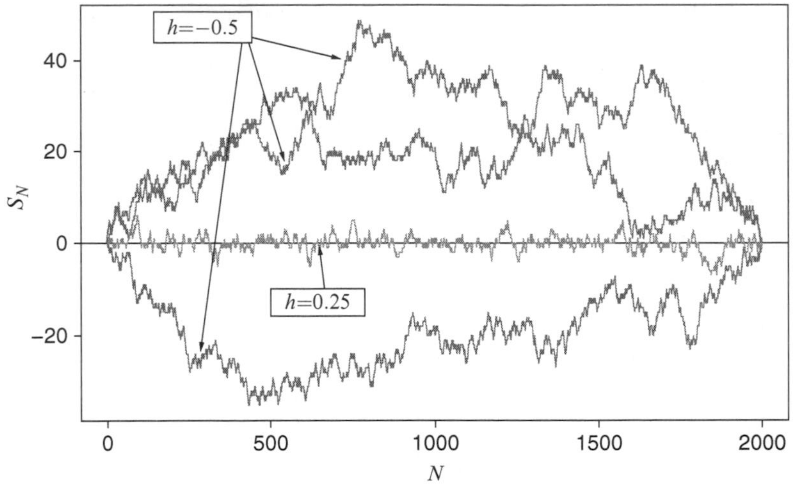

Fig. 2.2 Three trajectories for $h = -0.5$ and $N = 2{,}000$ (the underlying walk has $p = q = 1/2$) and one for $h = 0.25$. While the image clearly suggests that the first three trajectories are *delocalized*, i.e. they keep away from 0, and have a Brownian scaling, the maximum of the other trajectory is $+6$ and the minimum is -7, so the path is essentially localized at 0

Proof. The result follows from Proposition 2.4 by *routine* arguments. We go quickly through them, leaving some of the details to the reader. Let us point out that, in a sense, Proposition 2.4 is a detour and everything in the end boils down to renewal function estimates, but developing in arguments using the partition function has several advantages, like making connection with more general cases.

For what concerns (2.21) the result for $n = 1$ is immediate, since the probability we want to compute is $\exp(h)K(N)/Z_{N,h} \sim (1-p)^2$. For $n = 2$ instead one observes that for any choice of a sequence of natural numbers $\{a(N)\}_N$ such that $1 \ll a(N) \ll N$ we have

$$
\begin{aligned}
\mathbf{P}_{N,h}(L_N(S)=2) &= \frac{\sum_{n=1}^{N} K(n)K(N-n)\exp(2h)}{Z_{N,h}} \\
&= \frac{\sum_{n=1}^{a(N)-1} \cdots + \sum_{n=a(N)}^{N-a(N)-1} \cdots + \sum_{n=N-a(N)}^{N} \cdots}{Z_{N,h}} \\
&= 2\exp(h)(1-\exp(h))^2(1+o(1)) \\
&\quad + c_{h,K(\cdot)}^{-1} e^{2h}(1+o(1))N^{3/2} \sum_{n=a(N)}^{N-a(N)-1} n^{-3/2}(N-n)^{-3/2} \\
&= 2\exp(h)(1-\exp(h))^2(1+o(1)) + O(a(N)^{-1/2}),
\end{aligned} \tag{2.24}
$$

which is the result we were looking for. The general case is just a straightforward generalization which is best done by first observing that the configurations that contain points far from both 0 and N are negligible:

$$\mathbf{P}_{N,h}\left(\tau\cap[L,N-L]\neq\emptyset\right)\leq\frac{\sum_{n=L}^{N-L}Z_{n,h}Z_{N-n,h}}{Z_{N,h}}$$

$$\leq c_1\frac{\sum_{n=L}^{N-L}K(n)K(N-n)}{K(N)}\leq c_2\overline{K}(L),\quad(2.25)$$

where c_1 and c_2 are suitable (h and $K(\cdot)$ dependent) positive constants. In words, the result is simply saying that in the limit the process comes back a finite number of times close to 0 (each time it attempts to come back it has a probability $1-\exp(h)$ of not making it) and the behavior near N is just mirror symmetric (in law).

A proof of (2.22) is an immediate consequence of the fact that for $h>0$ we have $\mathrm{F}''(h)=\lim_N N^{-1}\mathrm{var}_{\mathbf{P}_{N,h}}(L_N(S))$, but this requires some work. So we take the cheaper path of observing that $\mathbf{P}_{N,h}$ actually coincides with the law of the $\widetilde{K}_h(\cdot)$-renewal conditioned to visit N: for every n and every $s\in\mathbb{N}^n$ such that $0<s_1<\ldots<s_n=N$ we have

$$\begin{aligned}\mathbf{P}_{N,h}\left(\tau\cap(0,N]=\{s_1,\ldots,s_n\}\right)&=\frac{e^{nh}K(s_1)K(s_2-s_1)\ldots K(N-s_{n-1})}{e^{N\mathrm{F}(h)}\mathbf{P}\left(N\in\widetilde{\tau}^{(h)}\right)}\\&=\frac{\widetilde{K}_h(s_1)\widetilde{K}_h(s_2-s_1)\ldots\widetilde{K}_h(N-s_{n-1})}{\mathbf{P}\left(N\in\widetilde{\tau}^{(h)}\right)}\\&=\mathbf{P}\left(\widetilde{\tau}^{(h)}\cap(0,N]=\{s_1,\ldots,s_n\}\,\middle|\,N\in\widetilde{\tau}^{(h)}\right).\end{aligned}\quad(2.26)$$

But since, by the law of large numbers, $\widetilde{\tau}_j^{(h)}/j$ tends as $j\to\infty$ to $\mathbf{E}[\widetilde{\tau}_1^{(h)}]$ almost surely, one directly obtains that $N^{-1}|\widetilde{\tau}^{(h)}\cap(0,N]|\longrightarrow 1/\mathbf{E}[\widetilde{\tau}_1^{(h)}]$ almost surely. Since the event $N\in\widetilde{\tau}^{(h)}$ has a probability bounded away from zero, for any sequence of events A_N such that $\mathbf{P}(A_N)\longrightarrow 0$, we have also $\mathbf{P}(A_N|N\in\widetilde{\tau}^{(h)})\longrightarrow 0$. Therefore, by using $A_N=\{|N^{-1}|\widetilde{\tau}^{(h)}\cap(0,N]|-1/\mathbf{E}[\widetilde{\tau}_1^{(h)}]|>\varepsilon\}$, we get (2.22).

For what concerns (2.23) consider first the case $t_n=t_\star$ and write much like for (2.26) (with $t_0:=0$ and assuming N larger than $t_\star$)

$$\begin{aligned}&\mathbf{P}_{N,h}\left(\tau\cap(0,t_n]=\{t_1,t_2,\ldots,t_n\}\right)=\\&\exp(hn)\left(\prod_{j=1}^n K(t_j-t_{j-1})\right)\frac{Z_{N-t_n,h}}{Z_{N,h}}=\prod_{j=1}^n\widetilde{K}_h(t_j-t_{j-1})\frac{Z_{N-t_n,h}\exp(\mathrm{F}(h)t_n)}{Z_{N,h}}\\&=\mathbf{P}\left(\widetilde{\tau}^{(h)}\cap(0,t_n]=\{t_1,t_2,\ldots,t_n\}\right)\left[\frac{\mathbf{P}\left(N-t_n\in\widetilde{\tau}^{(h)}\right)}{\mathbf{P}\left(N\in\widetilde{\tau}^{(h)}\right)}\right],\end{aligned}\quad(2.27)$$

where we have applied the renewal property and (2.13). But the term between the square brackets tends to 1 as N tends to infinity (because of the Renewal Theorem if $h > 0$, and because, if $h < 0$, partition functions coincide with renewal functions to which (2.15) applies). The case $t_\star > t_n$ of (2.23) can be dealt with by decomposing the probability according to the values of the first contact site t larger than $t_\star$ and by applying (2.27), that is by writing

$$\mathbf{P}_{N,h}\left(\tau\cap(0,t_\star]=\{t_1,\ldots,t_n\}\right) = \sum_{t=t_\star+1}^{N} \mathbf{P}_{N,h}\left(\tau\cap(0,t]=\{t_1,\ldots,t_n,t\}\right). \tag{2.28}$$

Actually at this stage one can for example use the argument used in (2.25) to restrict the summation only to values of t that are either close to $t_\star$ or close to N and then apply (2.15) and (2.27). By performing the summation we recover (2.23). □

2.2 The General Homogeneous Pinning Model

The asymptotic arguments that we have developed up to here essentially rely only on the fact that the tail distribution of the first return to the origin of the random walk S has a power law decay with exponent $3/2$. The first generalization that comes to mind is, possibly, considering higher dimensional random walks. These cases can be treated precisely along the same line, in fact one can show (see e.g. [22, Appendix A.6]) that if the increment of the random walk is a ($\mathbb{Z}^d$-valued) centered random variable with finite variance σ^2 (and $\mathbf{P}(S_1=0)\in(0,1)$ to avoid periodicity and triviality) then

$$K(n) \overset{n\to\infty}{\sim} c_d(\sigma^2)\times\begin{cases} 1/\left(n(\log n)^2\right) & \text{if } d=2 \\ 1/n^{1+|(d/2)-1|} & \text{if } d=1,3,4\ldots \end{cases} \tag{2.29}$$

with $c_d(\sigma^2)>0$. Another important fact is that $\sum_n K(n)=1$ if $d=1$ and 2, but $\sum_n K(n)<1$ for $d=3,4,\ldots$ But since our model in the end depends only on the inter-arrival law it is very natural to look at the renewal process τ as the basic underlying process (the *free process*) and put conditions on it: much of the literature has been in fact developed for $K(\cdot)$ *regularly varying* and it is possibly also natural to look at the case is which $\overline{K}(\cdot)$ ($\overline{K}(n)=\sum_{j>n}K(j)$, $n=0,1,\ldots$) is regularly varying. In order to make our arguments lighter we will consider a subclass of regularly varying inter-arrival laws supported on $\mathbb{N}$, that is we will assume that there exists $\alpha>0$ such that

$$K(n) \overset{n\to\infty}{\sim} \frac{\mathtt{c}_K}{n^{1+\alpha}} \quad \text{and} \quad K(n)>0 \ \text{ for } n\in\mathbb{N}. \tag{2.30}$$

The positivity condition can be relaxed at the expense of a series of tedious remarks that we spare to the reader (of course $K(n)\sim \mathtt{c}_K/n^{1+\alpha}$ implies $K(n)>0$ for n

sufficiently large). The choice of restricting to *trivial* regularly varying behavior (pure power law) is instead more substantial, above all because it excludes from our analysis the $d = 2$ case (2.29) and, more generally, the $\alpha = 0$ case, in which an interesting phenomenon does happen. But the gain in simplicity of exposition is considerable.

Note that we have not assumed $\sum_n K(n) = 1$: in general we set (again) $K(\infty) := 1 - \sum_{n=1}^{\infty} K(n)$ and we stress that $\sum_{n=1}^{\infty} \cdots$ means $\sum_{n \in \mathbb{N}} \cdots$. Let us write down explicitly the model:

$$\frac{\mathrm{d}\mathbf{P}_{N,h}}{\mathrm{d}\mathbf{P}}(\tau) = \frac{1}{Z_{N,h}} \exp\left(h\,|\tau \cap (0,N]|\right) \mathbf{1}_{N\in\tau}\,. \tag{2.31}$$

As we have already stressed, τ can be terminating or persistent and the following remark, that is going to be repeated in the most general context later, turns out to be quite helpful.

Remark 2.6. If τ is terminating, then the model is equivalent on events that depend on $\tau \cap (0,N]$ to the model based on the persistent $\widetilde{\tau}$ renewal with inter-arrival law $n \mapsto \widetilde{K}(n) := K(n)/(1-K(\infty))$ and h replaced by $h + \log(1-K(\infty))$. This can be easily verified by writing explicitly the probability of the event $\tau \cap (0,N] = \{t_1, t_2, \ldots, t_n\}$, $n \in \{1, \ldots, N\}$ and $0 < t_1 < t_2 < \ldots < t_n = N$. In particular, the two partition functions coincide (we are talking of $Z_{N,h}$ not of $Z_{N,h}^{\mathrm{f}}$!). This allows us to restrict in most of the cases our attention to the case in which the underlying renewal is persistent.

The generalization of Proposition 2.4 is in a sense straightforward, but it does present some novelties both from the viewpoint of mathematical tools (in fact: renewal theory estimates) and for the novel behaviors arising (see Fig 2.3).

Theorem 2.7. *For the partition function* $Z_{N,h} = \mathbf{E}[\exp(h\,|\tau \cap (0,N]|)\,;\, N \in \tau]$ *and the companion free partition function* $Z_{N,h}^{\mathrm{f}} = \mathbf{E}[\exp(h\,|\tau \cap (0,N]|)]$, *both based on the* $K(\cdot)$*-renewal, with* $K(\cdot)$ *as in* (2.30), *we have that*

$$\lim_{N\to\infty} \frac{1}{N} \log Z_{N,h} = \lim_{N\to\infty} \frac{1}{N} \log Z_{N,h}^{\mathrm{f}} = \mathrm{F}(h)\,, \tag{2.32}$$

where $\mathrm{F}(h)$ *– the free energy – is the unique solution of* (2.9) *if such a solution exists (that is if* $h \geq h_c := -\log(1-K(\infty))$*) and* $\mathrm{F}(h) := 0$ *otherwise. Moreover*

$$Z_{N,h} \overset{N\to\infty}{\sim} c_{h,K(\cdot)} \exp(\mathrm{F}(h)N) \times \begin{cases} N^0 & \text{if } h > h_c\,, \\ N^{\min(\alpha-1,0)} & \text{if } h = h_c \text{ and } \alpha \neq 1\,, \\ 1/\log N & \text{if } h = h_c \text{ and } \alpha = 1\,, \\ N^{-(1+\alpha)} & \text{if } h < h_c\,, \end{cases} \tag{2.33}$$

with the explicit value of the constant $c_{h,K(\cdot)} > 0$ *given below.*

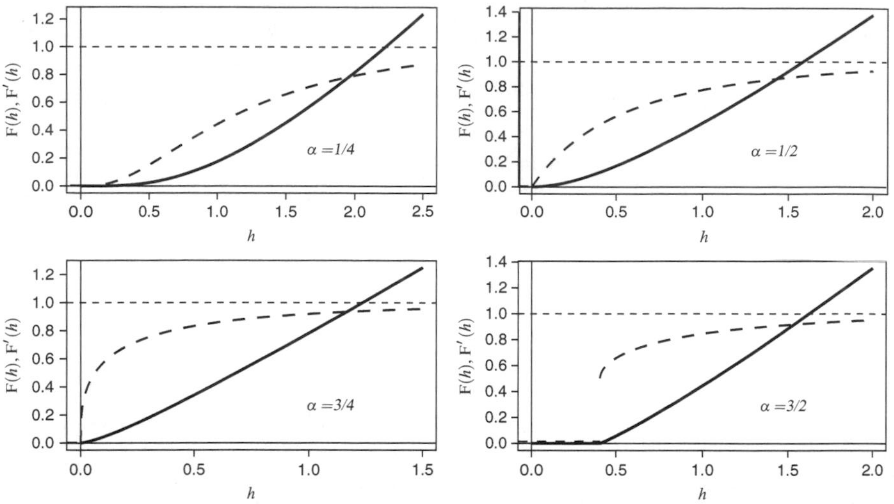

Fig. 2.3 Free energy ($\mathrm{F}(\cdot)$, *solid line*) and contact fraction ($\mathrm{F}'(\cdot)$, *dashed line*) for four values of α. The particular models we have chosen have $K(n) = \alpha\Gamma(n-\alpha)/(\Gamma(1-\alpha)n!) \overset{n\to\infty}{\sim} (\alpha/\Gamma(1-\alpha))n^{-1-\alpha}$ for $\alpha \in (0,1)$ and $K(n) = \Gamma(n-1/2)/(\sqrt{\pi}(n+1)!) \overset{n\to\infty}{\sim} (1/\sqrt{\pi})n^{-5/2}$ for the $\alpha = 3/2$ case. Such peculiar choices of $K(\cdot)$ are made because $\sum_n K(n)\exp(-\mathrm{F}n)$ can be made explicit by using By using the identity $\sum_{n=0}^{\infty} \Gamma(\beta+n)x^n/n! = \Gamma(\beta)(1-x)^{-\beta}$, that holds for $\beta \in \mathbb{R} \setminus \{0,-1,-2,\ldots\}$: for example when $\alpha \in (0,1)$ we have $\sum_n K(n)\exp(-\mathrm{F}n) = 1-(1-\exp(-\mathrm{F}))^{\alpha}$. In the first three cases $\sum_n K(n) = 1$ so that $h_c = 0$, but for $\alpha = 3/2$ we have $\sum_n K(n) = 2/3$ (nothing sacred about $2/3$, it is just an arbitrary choice!), so that $h_c = \log(3/2) = 0.405\ldots$. For the $\alpha = 1/2$ we have $K(n) = \mathbf{P}(\tau_1 = 2n)$, where τ the renewal set associated to the one dimensional symmetric simple random walk

Proof. All is of course in (2.13) [recall also (2.10)] and all we need are (sharp) renewal function estimates. These estimates are discussed at length in Appendix A, here we just recall the main results that can be summed up to: if $\widetilde{K}(\cdot)$ is an inter-arrival distribution (with $\widetilde{K}(\infty) := 1-\sum_n \widetilde{K}(n) \in (0,1)$) and $\widetilde{\tau}$ the corresponding renewal

1. If $\widetilde{K}(\infty) = 0$ and $\sum_n n\widetilde{K}(n) < \infty$ we have $\lim_{N\to\infty} \mathbf{P}(N \in \widetilde{\tau}) = 1/\sum_n n\widetilde{K}(n)$ (this is just the Renewal Theorem).
2. If $\widetilde{K}(\infty) = 0$ and $\widetilde{K}(n) \sim cn^{-1-\alpha}$ (with $c > 0$ and $\alpha \in (0,1)$) we have

$$\mathbf{P}(n \in \widetilde{\tau}) \overset{n\to\infty}{\sim} \frac{\alpha \sin(\pi\alpha)}{c\pi} n^{\alpha-1}. \tag{2.34}$$

3. If $\widetilde{K}(\infty) = 0$ and $\widetilde{K}(n) \sim c/n^2$ ($c > 0$) then

$$\mathbf{P}(n \in \widetilde{\tau}) \overset{n\to\infty}{\sim} \frac{1}{c\log n}. \tag{2.35}$$

4. If $\widetilde{K}(\infty) > 0$ and $\widetilde{K}(n) \sim cn^{-1-\alpha}$ (with $c > 0$ and $\alpha > 0$) then

$$\mathbf{P}(n \in \widetilde{\tau}) \overset{n\to\infty}{\sim} \frac{\widetilde{K}(n)}{\widetilde{K}(\infty)^2}. \tag{2.36}$$

Very little of the sharpness of these estimates is needed to establish (2.32) (see the proof of Proposition 2.4). Extracting (2.33) is instead a rather tedious book-keeping exercise with $\widetilde{K}(\cdot) = \widetilde{K}_h(\cdot)$ [cf. (2.10)]: let us go through it so that we determine $c_{h,K(\cdot)}$.

If $h > h_c$ we can apply point (1) and $\sum_n n\widetilde{K}_h(n) = 1/\mathrm{F}'(h)$ (use for example that the derivative of $\sum_n \widetilde{K}_h(n)$ with respect to h is zero), so that $c_{h,K(\cdot)} = \mathrm{F}'(h)$.

If $h < h_c$ we have $\widetilde{K}_h(\infty) > 0$ and we can apply point (4). The net result is that $c_{h,K(\cdot)} = \exp(h)/(1-\exp(h))^2$.

When $h = h_c$ instead notice first of all that $\widetilde{K}_h(\infty) = 0$, so $\widetilde{\tau}_h$ is persistent (regardless of the persistence properties of the reference renewal τ!). We distinguish the three cases $\alpha > 1$, $\alpha = 1$ and $\alpha < 1$. If $\alpha > 1$ we apply (1) and $c_{h,K(\cdot)} = \sum_n K(n)/\sum_n nK(n)$ (notice that we have used our convention that $\sum_n \ldots$ does not include $n = \infty$, so that $\sum_n nK(n) < \infty$). If $\alpha = 1$ we apply (3) and $c_{h,K(\cdot)} = \sum_n K(n)/c_K$. If $\alpha \in (0,1)$ then (2) yields $c_{h,K(\cdot)} = (\alpha \sin(\pi\alpha)\sum_n K(n))/(c_K \pi)$. □

Remark 2.8. Extracting from (2.33) the sharp asymptotic behavior of $Z^{\mathrm{f}}_{N,h}$ is an even more tedious exercise. The result is however definitely instructive and not void of interest, both for the sequel and for the intuition. We do not want to make the exposition too heavy and we refer to [22, Chap. 2], but we point out that the fact that the constrained partition function $Z_{N,h}$ is invariant under the transformation $(\tau, h) \mapsto (\widetilde{\tau}, h + \log \sum_n K(n))$ of Remark 2.6, does not imply that also $Z^{\mathrm{f}}_{N,h}$ is invariant (in fact this is false and, in some cases, even the large N behavior is different).

Extracting path properties from Theorem 2.7 is an exercise: result and proof are absolutely parallel to Proposition 2.5.

Proposition 2.9. *If $h < h_c = -\log \sum_n K(n)$ then for every $n \in \mathbb{N}$*

$$\lim_{N\to\infty} \mathbf{P}_{N,h}(L_N(S) = n) = (1-\exp(h-h_c))^2 n \exp((h-h_c)(n-1)), \tag{2.37}$$

while if $h > h_c$ we have that for every $\varepsilon > 0$

$$\lim_{N\to\infty} \mathbf{P}_{N,h}\left(\left|\frac{|\tau \cap (0,N]|}{N} - \mathrm{F}'(h)\right| \geq \varepsilon\right) = 0. \tag{2.38}$$

Moreover for every h, every n, every $t_\star$ and every $t \in \mathbb{N}^n$ such that $0 < t_1 < t_2 < \ldots < t_n \leq t_\star$ we have

$$\lim_{N\to\infty} \mathbf{P}_{N,h}(\tau \cap (0,t_\star] = \{t_1,\ldots,t_n\}) = \mathbf{P}\left(\tau^{(h)} \cap (0,t_\star] = \{t_1,\ldots,t_n\}\right), \tag{2.39}$$

that is the sequence $\{\mathbf{P}_{N,h}\tau^{-1}\}_N$ *of measures on* $\mathscr{P}_0$ *(cf. Sect. A.1.4 of Appendix A) converges weakly to* $\mathbf{P}(\widetilde{\tau}^{(h)})^{-1}$.

2.3 Phase Transition and Critical Behavior

This section focuses of the behavior of $\mathrm{F}(h)$ close to h_c. In view of (2.32) and of Remark 2.6, we can develop the arguments in the persistent set-up, that is when $h_c = 0$. The crucial estimate, like in Sect. 2.1.3, is understanding the asymptotic behavior of $\sum_n K(n)\exp(nx)$ for $x \searrow 0$. So let us set:

$$\Psi(x) \overset{x>0}{:=} 1 - \sum_{n=1}^{\infty} K(n)\exp(-nx), \tag{2.40}$$

and let us compute:

$$\begin{aligned} 1 - \sum_{n=1}^{\infty} K(n)\exp(-nx) &= 1 - \sum_{n=1}^{\infty} \left(\overline{K}(n-1) - \overline{K}(n)\right)\exp(-nx) \\ &= (1-\exp(-x)) \sum_{n=0}^{\infty} \exp(-nx)\overline{K}(n)\,. \end{aligned} \tag{2.41}$$

Therefore, when $\alpha > 1$, one directly sees that $\Psi(x) \sim x\mathbf{E}[\tau_1]$ as $x \searrow 0$. If instead $\alpha \in (0,1)$, by Riemann sum approximation, one obtains

$$\Psi(x) \overset{x\searrow 0}{\sim} x\sum_n \frac{\mathrm{c}_K \exp(-xn)}{\alpha n^{\alpha}} \sim \frac{x^{\alpha}\mathrm{c}_K}{\alpha}\int_0^{\infty} t^{-\alpha}\exp(-t)\mathrm{d}t \sim \mathrm{c}_K \frac{\Gamma(1-\alpha)}{\alpha} x^{\alpha}\,. \tag{2.42}$$

For $\alpha = 1$ we set $\ell(n) := \sum_{j=1}^{n} 1/j$ for $n \in \mathbb{N}$ and $\ell(0) := 0$ so that

$$\frac{\Psi(x)}{\mathrm{c}_K} \overset{x\searrow 0}{\sim} x\sum_{n=1}^{\infty} \frac{\exp(-xn)}{n} = x(1-\exp(-x))\sum_{n=1}^{\infty} \ell(n)\exp(-xn)\,, \tag{2.43}$$

Now note that $\sum_{n=1}^{\infty} \ell(n)\exp(-xn) \sim \sum_n \log(n)\exp(-xn)$ and since we have that $\sum_n \log(xn)\exp(-xn)$ is $O(1/x)$, we see that $\sum_{n=1}^{\infty} \ell(n)\exp(-xn)$ is asymptotically equivalent to $\log(1/x)\sum_{n=1}^{\infty}\exp(-xn) \sim x^{-1}\log(1/x)$. Therefore if $\alpha = 1$

$$\Psi(x) \overset{x\searrow 0}{\sim} \mathrm{c}_K x\log(1/x)\,. \tag{2.44}$$

By recalling that $\Psi(\mathrm{F}(h)) = 1 - \exp(h)$, by inverting the asymptotic relations we obtain the behavior of $\mathrm{F}(h)$ for $h \searrow 0$: we sum up the result in the following statement.

Theorem 2.10. *For $K(\cdot)$ as in* (2.30)*, $\mathrm{F}(\cdot)$ as in Theorem 2.7 and h_c equal to $-\log \sum_n K(n)$, we have that*

$$\mathrm{F}(h) \overset{h \searrow h_c}{\sim} C(K(\cdot)) \begin{cases} h - h_c & \text{if } \alpha > 1\,, \\ (h-h_c)/\log(1/(h-h_c)) & \text{if } \alpha = 1\,, \\ (h-h_c)^{1/\alpha} & \text{if } \alpha \in (0,1)\,, \end{cases} \tag{2.45}$$

where

$$C(K(\cdot)) = \begin{cases} \sum_n K(n)/\sum_n nK(n) & \text{if } \alpha > 1\,, \\ 1/c_K & \text{if } \alpha = 1\,, \\ \left((\alpha \sum_n K(n))/(c_K \Gamma(1-\alpha))\right)^{1/\alpha} & \text{if } \alpha \in (0,1)\,. \end{cases} \tag{2.46}$$

If $\mathrm{F}(\cdot)$ is C^k (of course the issue is at h_c), then $\mathrm{F}(h) = o((h-h_c)^k)$ for $h \searrow h_c$. Therefore Theorem 2.10 directly implies that, for $k = 2, 3, \ldots$, $\mathrm{F}(\cdot)$ is not C^k for $\alpha \geq 1/k$. Moreover, it is not C^1 for $\alpha > 1$ (but of course it is C^0). By using the convexity of $\mathrm{F}(\cdot)$ one directly extracts also that, for $\alpha \leq 1$, $\mathrm{F}(\cdot)$ is C^1: since $\mathrm{F}'(\cdot)$ is non-decreasing (and well-defined except possible at h_c), a discontinuity at h_c of $\mathrm{F}'(\cdot)$ implies $\mathrm{F}(h) \geq c(h-h_c)$ for $h > h_c$, with $c = \lim_{h \searrow h_c} \mathrm{F}'(h) > 0$, which contradicts the estimate in Theorem 2.10. In general one has instead to resort to a direct estimate (that can be found Appendix A, Theorem A.8). The net result is summed up in the next statement in which we use the standard terminology: a phase transition is a point of non-analiticity of the free energy, this point is called *critical*, and the phase transition is said *of kth order* ($k \in \mathbb{N}$) if the free energy is, at the critical point, C^{k-1}, but not C^k.

Proposition 2.11. *The homogeneous pinning model* (2.31) *has a phase transition of kth order, $k = 2, 3, \ldots$, at $h = h_c$ if $\alpha \in [1/k, 1/(k-1))$. The transition is of second order also if $\alpha = 1$, while it is of first order for $\alpha > 1$.*

2.4 A First Look at a Crucial Notion: The Correlation Length

The notion of correlation length plays a central role in the study of statistical mechanics systems. In general, even for a given system there are plenty of reasonable definitions of correlation length. Let us see for the homogeneous pinning system: we have seen (Proposition 2.5) that the $N \to \infty$ model is the renewal with inter-arrival law $\widetilde{K}_h(\cdot)$. In this case the first notion of correlation length that comes to mind is given by looking at the correlation

$$\mathbf{E}\left[\widetilde{\delta}_m \widetilde{\delta}_{m+n}\right] - \mathbf{E}\left[\widetilde{\delta}_m\right] \mathbf{E}\left[\widetilde{\delta}_{m+n}\right] = \mathbf{E}\left[\widetilde{\delta}_m\right] \left(\mathbf{E}\left[\widetilde{\delta}_n\right] - \mathbf{E}\left[\widetilde{\delta}_{m+n}\right]\right), \tag{2.47}$$

where $\widetilde{\delta}_n = \mathbf{1}_{n\in\widetilde{\tau}^{(h)}}$. If $h \le h_c$ (delocalized regime) then, thanks to Theorem A.2 and to Theorem A.4, one directly sees that for every m the correlation decays, as $n \to \infty$, with a power law: since the correlation length is naturally defined as the reciprocal of the exponential decay rate of the correlations, we see that in this case the correlation length is ∞. If instead $h > h_c$ one can, for the sake of simplicity, take the limit $m \to \infty$, so that one is effectively talking about the covariance of the stationary renewal: by the Renewal Theorem, applied to (2.47), the correlation length this time is read off

$$\mathbf{E}\left[\widetilde{\delta}_n\right] - \frac{1}{\widetilde{\mu}(h)}, \quad \text{with } \widetilde{\mu}(h) := \mathbf{E}\widetilde{\tau}_1^{(h)}. \tag{2.48}$$

The correlation length is therefore given by the reciprocal of the rate of convergence of the renewal function to its asymptotic value. The renewal equation comes to our help in order to compute it, but things are not as easy as one might think at first. It is a standard result [28] that if the inter-arrival law decays exponentially (more precisely: in the case of a recurrent $\widetilde{K}(\cdot)$-renewal such that $\sup_n \exp(cn)\widetilde{K}(n) < \infty$ for some $c > 0$), then the renewal function converges to its limit exponentially fast. As it is well known since a long time (see for example [26]) however, the relation between the rate of decay of $\widetilde{K}(\cdot)$ and the one of the renewal function are in general rather *unrelated* (see [23] for examples and several references). But what is going to be important for our discussion is that, in the context we consider, one can establish a general result. Namely that

Proposition 2.12. *[23] Choose an inter-arrival law $K(\cdot)$ that satisfies* (2.30)*. Then there exists $h_0 \in (h_c, \infty]$ such that for $h \in (h_c, h_0)$ we have*

$$\mathbf{P}\left(n \in \widetilde{\tau}^{(h)}\right) - \frac{1}{\widetilde{\mu}(h)} \stackrel{n\to\infty}{\sim} c(h)K(n)\exp(-\mathrm{F}(h)n), \tag{2.49}$$

with $c(h)$ a positive (explicit) constant. So, in particular, we have

$$\lim_{n\to\infty} -\frac{1}{n}\log\left(\mathbf{P}\left(n \in \widetilde{\tau}^{(h)}\right) - \frac{1}{\widetilde{\mu}(h)}\right) = \mathrm{F}(h). \tag{2.50}$$

This result can be read as saying that the correlation length $\kappa = \kappa(h)$ is equal to $1/\mathrm{F}(h)$, at least when the system is close to criticality. The fact that (in general) we can link the correlation length to the free energy only close to criticality is not a problem because the correlation length becomes important precisely close to criticality, that is when it diverges.

The role of $\kappa(h)$ emerges clearly also from (2.13): the exponential growth of the partition function sets in when N is about $\kappa(h)$ (look at Fig. 2.1). Figure 2.1 definitely suggests another correlation length: $\widetilde{\kappa}(h) := \inf\{N : Z_{N,h} > 1\}$, where the value 1 is a bit arbitrary (at this stage), but it is a natural reference point. Why this is not such a bad definition will be clear later on: for the moment we register the fact that $\log\kappa(h) \sim \log\widetilde{\kappa}(h)$ as $h \searrow h_c$.

2.5 Why Do People Look at Pinning Models? A Modeling Intermezzo

The main purpose of these notes is to investigate the effect of disorder on statistical mechanics models, notably on phase transitions and critical phenomena. Pinning models turn out to be a particularly favorable context to attack this daunting issue.

But pinning models have received widespread attention, notably in physics, chemistry and biology. Let us have a quick look at this direction by considering three different instances: this will serve also to motivate the introduction of *disorder*.

2.5.1 *Polymer Pinning by a Defect*

Polymers are chains of repetitive units (monomers) that may or may not be identical. Polymer modeling is tightly related to random walks in the sense that the most basic model of a polymer is the random walk. Less simplistic models include a self-avoiding constraint and/or increment correlation. In addition, polymers are often in interaction with an environment: the presence in the environment of an attractive (or repulsive) region may have a substantial effect on the polymer trajectory. When such a region is a point or a line (but it could also be a plane or a hyperplane) then a natural basic model, in which we either disregard self-avoidance or we implement it by looking at the so-called *directed polymers*, is precisely the pinning model. For more on this, see [22, Chap. 1] and references therein.

2.5.2 *Interfaces in Two Dimensions*

There is very deep link between interfaces in two dimensional (discrete spin) models and random walks (e.g. [1]). It is sketched in Fig. 2.4, both in the case in which the arising walk is free and when there is a pinning effect. The figure is based on the Ising model that we introduce also because it comes up later on in these notes. An Ising model in the rectangular box $\Lambda = \prod_{i=1}^{d}(-L_i, L_i) \cap \mathbb{Z}^d$ ($L_i \in \mathbb{N}$) is a measure on $\Omega_\Lambda := \{-1,+1\}^{\Lambda \cup \partial\Lambda}$, where $\partial\Lambda$ is the external boundary of Λ, that is $\partial\Lambda = \{x \in \mathbb{Z}^d \setminus \Lambda : |x-y| = 1 \text{ for a } y \in \Lambda\}$. The measure is determined once we fix a value $\beta \geq 0$ (the *inverse temperature*) and an element $\eta \in \{-1,+1\}^{\partial\Lambda}$, the boundary condition, and then we say that the probability $\mu_{\Lambda,\eta}(\sigma)$ of observing the configuration $\sigma \in \Omega_\Lambda$ is

$$\mu_{\Lambda,\eta}(\sigma) = \begin{cases} \exp(-\beta H_\Lambda(\sigma))/Z_{\Lambda,\beta,\eta} & \text{if } \sigma(x) = \eta(x) \text{ for } x \in \partial\Lambda\,, \\ 0 & \text{otherwise}\,, \end{cases} \tag{2.51}$$

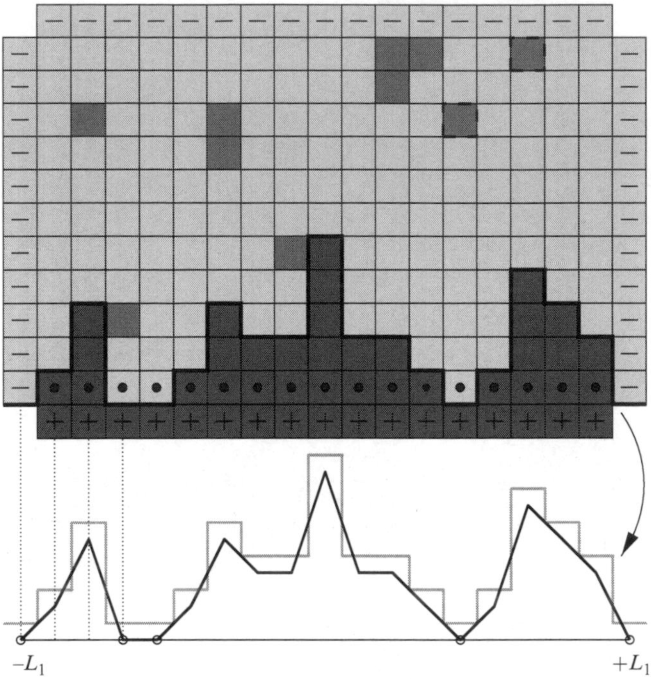

Fig. 2.4 We are drawing a configuration of the two-dimensional Ising model in the finite box with $L_1 = 9$, $L_2 = 6$ and boundary conditions that are $+1$ on $\{x : x_2 = -L_2, x_1 = -L_1 + 1, \ldots, L_1 - 1\}$ and -1 on the rest of $\partial^+\Lambda$. The up spins $(+)$ are identified by a *gray* or a *dark gray square*, while the down spins $(-)$ are *light gray*. The spins on which a boundary magnetic field acts are marked by *large black dots*. In the text, to which we refer for details, it is sketched the explanation of the reduction of the spin configuration to an *interface*: the interface is reproduced below the spin configuration with an equivalent but more natural representation if we look at it as a random walk path

where $Z_{\Lambda,\beta,\eta}$ is the normalization constant and $H_\Lambda(\sigma)$ is the energy of the model that we choose to be

$$H_\Lambda(\sigma) = -\frac{1}{2}\sum_{x,y} J(x,y)\sigma(x)\sigma(y) - \sum_x h(x)\sigma(x), \tag{2.52}$$

where the sums are over $\Lambda \cup \partial\Lambda$ and $J(x,y) = 0$ unless $|x-y| = 1$. Of course the most basic Ising model is the one in which $J(x,y) = 1$ for every $|x-y| = 1$ and $h(x)$ does not depend on x, but the general case here serves two purposes:

1. Later on we will discuss the generalization of what we develop for pinning models to more general cases and the disordered Ising model, that is the case in which $\{J(x,y)\}_{x,y}$ and/or $\{h(x)\}_x$ are realizations of families of IID random variables, is at the heart of the progress in the statistical mechanics of disordered

systems, much as the non-disordered Ising model is at the heart of the progress in statistical mechanics.

2. The *interface line*, or *phase separation line*, reduces to a random walk in the strongly anisotropic limit and a suitable choice of $h(\cdot)$ has the effect of a pinning potential: this is what we are going to explain next.

In (the upper part of) Fig. 2.4 we draw a configuration of the two dimensional Ising model in a finite box Λ: spins are drawn in small boxes that are either light gray, gray or dark gray. We are thinking of the case in which $J(x,y) = 0$ unless $|x-y| = 1$ and $J(x,y) = J_1 \geq 0$ (respectively $J(x,y) = J_2 \geq 0$) if $x-y = (\pm 1, 0)$ (respectively $x-y = (0, \pm 1)$). We also restrict our attention to $h(\cdot) \equiv 0$ for the moment, that is, we think of an Ising model without external (magnetic) field and with nearest neighbor interactions that can be different along the horizontal and vertical directions (the infinite volume limit of this model has been solved by Lars Onsager [4], a result that has deeply marked statistical mechanics).

As we are trying to convince the reader with the figure, such a spin configuration can be mapped to a set of contours (this is a very classical construction, see e.g. [18, Chap. 2]). All the contours are closed lines, except one that goes from the lower left corner to the lower right corner: we call such an open contour *interface* and we stress that the existence of an open contour is directly related to the boundary conditions (for example: all spins $+1$ on the boundary entails all contours are closed). Note that in the limit $J_1 \to \infty$ the configuration we have drawn has probability zero: a positive probability configuration is rather the one in which we switch to -1 all spins in the gray squares and in this case only one contour *survives*: the interface. More is true: in this limit the interface is a trajectory of a random walk with increments in $\mathbb{Z}$, starting in the lower left corner and ending on the lower right corner. As a matter of fact, it is an easy exercise to show that the law of such an Ising interface is just the law of the walk we have just mentioned (to be precise, the probability that the increment is equal to n is $const.\exp(-\beta J_2 |n|)$) conditioned not to exit the box Λ. If now we consider a very tall box ($L_2 \to \infty$) we are just dealing with a random walk bridge constrained not to go below the height of its starting (and arrival) point. The lower part of the figure draws the interface with a slightly different convention that has the advantage to be closer to the customary way of drawing random walk trajectories.

If now we allow what is usually called a *boundary magnetic field*, that is if we set for example $h((i, -L_2+2)) = -h < 0$ for $i = -L_1+1, \ldots, L_1-1$, spins of value -1 are favored in the sites on which we have put the field (the sites on which we put the boundary field are marked by large black dots). What is the effect of the boundary field on the random walk trajectory? The answer is simply that there is a reward of $h > 0$ for the walk to stick to the bottom line: we are therefore just dealing with a homogeneous pinning model.

Two remarks to close this issue are in order.

1. Of course all that we have discussed becomes more delicate and definitely not as straightforward if $J_1 < \infty$, nonetheless the simplified $J_1 = \infty$ case to a certain

extent turns out not to be an oversimplification (see e.g. [37] and references therein).

2. It is of course natural to choose $h(i, -L_2 + 2)$ with a non-trivial dependence on i (for example, we could choose them by coin tossing $(\pm h)$): the arising random walk model is a inhomogeneous pinning model that fits (and motivates!) the definition of inhomogeneous models of the next chapters.

2.5.3 DNA Denaturation: The Poland–Scheraga Model

Understanding the very complex geometrical structure in which two complementary DNA strands (two polymers) are found in cell nuclei is a long standing issue on which a lot of effort is invested. There are of course plenty of issues: we focus just on the fact that two complementary strands are not tightly bind all the times (as a matter of fact, unbinding is necessary in particular for copying the genetic code to another polymer, the RNA) and unbinding, above all local unbinding, happens *all the time* as a standard consequence of thermal fluctuations. Biologists and physicists have developed models for such a phenomenon and a basic, but apparently rather effective model, even on a quantitative level, is just based on pinning models [20]. Starting off in the most naive way we can model two-stranded DNA by two directed walks interacting via pinning potentials, see e.g. [30] and references therein. Since the difference of two independent random walks is still a random walk, we are dealing with a standard pinning model. However directed walk models lead to values of α that are in contrast with observations. In fact the three dimensional model, that corresponds to the walk in two dimensions plus the *fixed direction*, yields $\alpha = 0$, a case not treated in this notes that however leads to a C^∞ behavior of the free energy [22, Chap. 2], while there is a tendency to believe that the transition is first order, even if such a statement has to be taken with caution because real DNA experiments are not about infinitely long strands, see for example [6, 27]. To make a long story short, the bio-physical community seems to have settled that renewal pinning models with $\alpha \approx 1.15$ is a reasonably good model for DNA denaturation [6, 15]: however what is most important is that inhomogeneous interactions need to be taken into account, unless one is dealing with synthetic DNA made up by one strand containing only *Adenine* (respectively *Cytosine*) bases and the the other strand containing only *Thymine* (respectively *Guanine*) bases. A few more details can be found in Fig. 2.5 and its caption.

2.6 A Look at the Literature

Much of this chapter is devoted to the homogeneous pinning model. This model, at least in some random walk cases, has been the object of several works in the physical literature at the beginning of the 1980s proposing *different* exact solutions (e.g. [7, 29]), but the generalized model and a comprehensive view identifying the

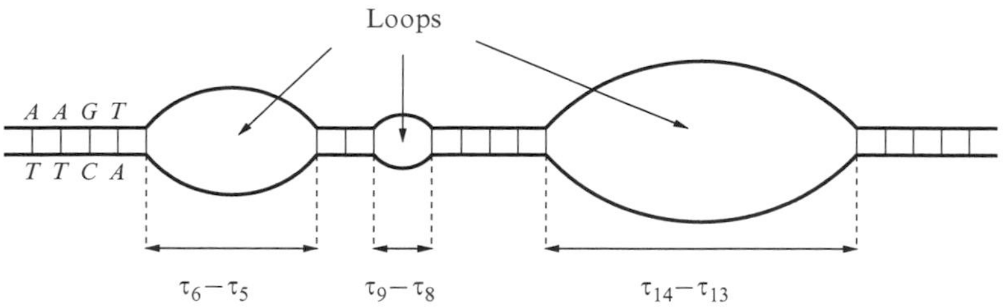

Fig. 2.5 A schematic, standard, view of DNA denaturation. The two *thick lines* are the DNA strands. They may be paired, gaining thus energetic contributions that depend on whether the base pair is A–T or G–C (the model is therefore inhomogeneous: A–T bonds are weaker than G–C bonds). The sections of unpaired bases are called *loops*. The DNA portion in the drawing corresponds to the renewal model trajectory with $\tau_j - \tau_{j-1} = 1$ except for three inter-arrivals (so loops correspond to inter-arrivals of length 2 or more)

general mechanism behind the *various exact solutions* are due to Fisher [19]. The approach given here, however, is not the one in [19], that goes by computing the series $\sum_N z^N Z_{N,h}$. We aim directly at $Z_{N,h}$ and at its interpretation in term of renewal function: this approach has been developed in [25, Appendix A] and [10]. It has been proven useful also beyond the renewal set-up, notably for Markov renewal processes that cover a very wide class of models: pinning and copolymers in periodic environments [11], pinning of directed semi-flexible polymers [9], pinning on layered interfaces [12] and pinning of random walks with continuous increments [33] (the Brownian motion case has been treated for example in [14, 31] by different techniques and we refer to [22] for further references on the vast literature on homogeneous pinning).

While of course the modeling aspects must not be neglected, the approach given here shows that the physical solution of the homogeneous pinning model (notably, free energy estimates) are just a subset of *classical* renewal theory developed in the 1950s and 1960s (e.g. [16, 17, 21], see [3] for further references).

Further considerations and references on path properties of the limit ($N \to \infty$) process can be found in [22, Chap. 2], notably the scaling limits at criticality that makes a link with the theory of *regenerative sets and subordinators* [5].

Dynamical issues have been left completely out and that will not change in the next chapters: these notes are about the equilibrium measure, but the dynamical issues are of great interest (see for example [8]).

In [2] the author provides a class of random walks with increments taking values ± 1 that have regularly varying return time distribution: therefore this work exhibits walks for which (2.30) holds, for arbitrary α.

Section 2.4 introduces the notion of correlation length: it is difficult to stress enough the role of such a concept in statistical mechanics. But it is also difficult to treat it in a satisfactory way without mentioning the *disordered* case: for here we content ourselves with adding the references [24, 34–36], that deal in part with the homogeneous case (and are very relevant for the disordered case), and [32] that develops a mathematical viewpoint on the *finite size scaling* properties of the homogeneous models, that is on the behavior of the system of correlation length size, close to criticality (when the correlation length diverges).

References

1. D.B. Abraham, *Surface structures and phase transitions, exact results*, in *Phase Transitions and Critical Phenomena*, vol. 10 (Academic, London, 1986), pp. 1–74
2. K.S. Alexander, Excursions and local limit theorems for Bessel-like random walks. Electron. J. Probab. **16**, 1–44 (2011)
3. S. Asmussen, *Applied Probability and Queues*, 2nd edn. (Springer, New York, 2003)
4. R.J. Baxter, *Exactly solved models in statistical mechanics* (Academic, London, 1982)
5. J. Bertoin, *Subordinators: examples and applications*. Lectures on probability theory and statistics (Saint-Flour, 1997), Lecture Notes in Mathematics, vol. 1717 (1999), pp. 1–91
6. R. Blossey, E. Carlon, Reparametrizing the loop entropy weights: effect on DNA melting curves. Phys. Rev. E **68**, 061911 (2003) (8 pages)
7. T.W. Burkhardt, Localization-delocalization transition in a solid-on-solid model with a pinning potential. J. Phys. A Math. Gen. **14**, L63–L68 (1981)
8. P. Caputo, F. Martinelli, F.L. Toninelli, On the approach to equilibrium for a polymer with adsorption and repulsion. Electron. J. Probab. **13**, 213–258 (2008)
9. F. Caravenna, J.-D. Deuschel, Pinning and wetting transition for (1+1)-dimensional fields with Laplacian interaction. Ann. Probab. **36**, 2388–2433 (2008)
10. F. Caravenna, G. Giacomin, L. Zambotti, Sharp asymptotic behavior for wetting models in (1+1)-dimension. Electron. J. Probab. **11**, 345–362 (2006)
11. F. Caravenna, G. Giacomin, L. Zambotti, A renewal theory approach to periodic copolymers with adsorption. Ann. Appl. Probab. **17**, 1362–1398 (2007)
12. F. Caravenna, N. Pétrélis, A polymer in a multi-interface medium. Ann. Appl. Probab. **19**, 1803–1839 (2009)
13. J.B. Conway, *Functions of One Complex Variable*, 2nd edn. Graduate Texts in Mathematics, vol. 11 (Springer, New York, 1978)
14. M. Cranston, L. Koralov, S. Molchanov, B. Vainberg, Continuous model for homopolymers. J. Funct. Anal. **256**, 2656–2696 (2009)
15. D. Cule, T. Hwa, Denaturation of heterogeneous DNA. Phys. Rev. Lett. **79**, 2375–2378 (1997)
16. W. Feller, *An Introduction to Probability Theory and Its Applications*, vol. I, 3rd edn. (Wiley, New York, 1968)
17. W. Feller, *An Introduction to Probability Theory and Its Applications*, vol. II, 2nd edn. (Wiley, New York, 1971)
18. R. Fernández, J. Frölich, A.D. Sokal, *Random Walks, Critical Phenomena, and Triviality in Quantum Field Theory*. Texts and Monographs in Physics (Springer, New York, 1992)
19. M.E. Fisher, Walks, walls, wetting, and melting. J. Stat. Phys. **34**, 667–729 (1984)
20. M. Fixman, J.J. Freire, Theory of DNA melting curves. Biopolymers **16**, 2693–2704 (1977)
21. A. Garsia, J. Lamperti, A discrete renewal theorem with infinite mean. Comment. Math. Helv. **37**, 221–234 (1963)
22. G. Giacomin, *Random Polymer Models* (Imperial College Press, London, 2007)
23. G. Giacomin, Renewal convergence rates and correlation decay for homogeneous pinning models. Electron. J. Probab. **13**, 513–529 (2008)
24. G. Giacomin, *Renewal sequences, disordered potentials, and pinning phenomena*, in Spin Glasses: Statics and Dynamics, Progress in Probability, vol. 62 (2009), pp. 235–270
25. G. Giacomin, F.L. Toninelli, Smoothing effect of quenched disorder on polymer depinning transitions. Commun. Math. Phys. **266**, 1–16 (2006)

26. C.R. Heathcote, Complete exponential convergence and some related topics. J. Appl. Probab., **4**, 217–256 (1967)
27. Y. Kafri, D. Mukamel, L. Peliti, Why is the DNA denaturation transition first order? Phys. Rev. Lett. **85**, 4988–4991 (2000)
28. D.G. Kendall, *Unitary dilations of Markov transition operators and the corresponding integral representation of transition probability matrices*, in Probability and Statistics, ed. by U. Grenander (Almqvist and Wiksell, Stockholm, 1959), pp. 138–161
29. J.M.J. van Leeuwen, H.J. Hilhorst, Pinning of rough interface by an external potential. Phys. A **107**, 319–329 (1981)
30. D. Marenduzzo, A. Trovato, A. Maritan, Phase diagram of force-induced DNA unzipping in exactly solvable models. Phys. Rev. E **64**, 031901 (2001) (12 pages)
31. B. Roynette, M. Yor, *Penalising Brownian Paths*. Lecture Notes in Mathematics, vol. 1969 (Springer, New York, 2009)
32. J. Sohier, Finite size scaling for homogeneous pinning models. ALEA Lat. Am. J. Probab. Math. Stat. **6**, 163–177 (2009)
33. J. Sohier, Phénomènes d'accrochage et théorie des fluctuations, PhD thesis, Univ. Paris Diderot, November 2010
34. F.L. Toninelli, Critical properties and finite-size estimates for the depinning transition of directed random polymers. J. Stat. Phys. **126**, 1025–1044 (2007)
35. F.L. Toninelli, Correlation lengths for random polymer models and for some renewal sequences. Electron. J. Probab. **12**, 613–636 (2007)
36. F.L. Toninelli, *Localization transition in disordered pinning models. Effect of randomness on the critical properties*, in Methods of Contemporary Mathematical Statistical Physics, Lecture Notes in Mathematics, vol. 1970, 129–176 (2009)
37. Y. Velenik, Localization and delocalization of random interfaces. Probab. Surv. **3**, 112–169 (2006)

Chapter 3
Introduction to Disordered Pinning Models

Abstract We introduce the disorder disordered version of the pinning models, both in their quenched and annealed version. We define the free energy of the model and show that also in this case a localization/delocalization transition takes place. Most of the results presented in this chapter may be considered as *soft*, but they are the result of a subtle, albeit possibly standard in statistical mechanics, way of combining convexity and super-additivity properties. These techniques are repeatedly used in the sequel of these notes.

3.1 The Disordered Pinning Model

We are going now to focus on the inhomogeneous model

$$\frac{\mathrm{d}\mathbf{P}_{N,\omega}}{\mathrm{d}\mathbf{P}}(\tau) = \frac{1}{Z_{N,\omega}} \exp\left(\sum_{n=1}^{N} (\beta\omega_n + h)\delta_n\right) \delta_N, \tag{3.1}$$

where $\delta_n := \mathbf{1}_{n\in\tau}$, $\beta \geq 0$, $h \in \mathbb{R}$ and $\omega = \{\omega_1, \omega_2, \ldots\}$ is a sequence of real numbers, that we call *charges*. The partition function $Z_{N,\omega}$ should rather be written as $Z_{N,\omega,\beta,h}$ but we will go for the lighter notation, unless (strictly) necessary (we also set $Z_{0,\omega} := 1$). Of course (3.1) directly generalizes (2.31) of Sect. 2.2, when $\beta > 0$ and the sequence ω is not trivial.

We will be interested in the case in which ω is the realization of an IID sequence of random variables (but, once again, we will switch from random variables to realization and back without changing notation). There are therefore random interactions, in fact random one-body potentials, that we call *quenched disorder*: the term *quenched* refers to the fact that we define the system for a typical realization of the charges (or disorder) ω.

G. Giacomin, *Disorder and Critical Phenomena Through Basic Probability Models*, Lecture Notes in Mathematics 2025, DOI 10.1007/978-3-642-21156-0_3,
© Springer-Verlag Berlin Heidelberg 2011

Definition 3.1. (General charge distribution.) The basic assumptions that we make on the IID sequence $\{\omega_n\}_{n\in\mathbb{N}}$ (or $\{\omega_n\}_{n\in\mathbb{Z}}$) of the charges are that

$$\mathrm{M}(t) := \mathbb{E}\left[\exp(t\omega_1)\right] < \infty, \tag{3.2}$$

for every $t \in \mathbb{R}$. We also assume, without loss of generality, that $\mathbb{E}[\omega_1] = 0$ and that $\mathbb{E}[\omega_1^2] = 1$.

We will tackle this model by focusing first on the properties of the partition function

$$Z_{N,\omega} = \mathbf{E}\left[\exp\left(\sum_{n=1}^{N}(\beta\omega_n + h)\delta_n\right)\delta_N\right], \tag{3.3}$$

that, this time, is a random variable. To be more precise we will focus on the free energy of the system

$$\mathrm{F}(\beta,h) := \lim_{N\to\infty}\mathrm{F}_N(\beta,h) \quad \text{with } \mathrm{F}_N(\beta,h) := \frac{1}{N}\mathbb{E}\log Z_{N,\omega}, \tag{3.4}$$

where the existence of the limit follows from the super-additive property of the sequence $\{\mathbb{E}\log Z_{N,\omega}\}_{N\in\mathbb{N}}$: for $M = 1,\ldots,N-1$ in fact we have (with $(\theta\omega)_n := \omega_{n+1}$)

$$\begin{aligned}
\log Z_{N,\omega} &\geq \log \mathbf{E}\left[\exp\left(\sum_{n=1}^{N}(\beta\omega_n + h)\delta_n\right)\delta_N\delta_M\right] \\
&= \log \mathbf{E}\left[\exp\left(\sum_{n=1}^{M}(\beta\omega_n + h)\delta_n\right)\delta_M\right] \\
&\quad + \log \mathbf{E}\left[\exp\left(\sum_{n=M+1}^{N}(\beta\omega_n + h)\delta_n\right)\delta_N \,\middle|\, \delta_M = 1\right] \\
&= \log Z_{M,\omega} + \log Z_{N-M,\theta^M\omega},
\end{aligned} \tag{3.5}$$

((3.5) becomes trivial for $M = 0$ and $M = N$, since we set $Z_{0,\omega} = 1$) so that for $M = 0,1,\ldots,N$

$$\mathbb{E}\log Z_{N,\omega} \geq \mathbb{E}\log Z_{M,\omega} + \mathbb{E}\log Z_{N-M,\omega}. \tag{3.6}$$

Let us also remark that $\mathbb{E}\log Z_{N,\omega} \leq \mathbb{E}\sum_{n=1}^{N}(\beta|\omega_n| + |h|) = O(N)$, so that the limit in (3.4) exists, it is finite and it coincides with the supremum of the sequence (see for example [6, Appendix A.7]

$$\mathrm{F}(\beta,h) = \sup_N \mathrm{F}_N(\beta,h), \tag{3.7}$$

but the reader may want to take it as a useful exercise).

The free energy will play a crucial role for what follows and we are going to disregard for a while the path properties of the process. Before doing that we just want to make an elementary observation that may help in understanding why $\mathbf{P}_{N,\omega}$ is (in general) non trivial. Let us place ourselves in the random walk set-up of the beginning of Chap. 2 (see Fig. 3.1) and let us consider the free case (analogous observations hold in the general set-up):

$$\frac{\mathrm{d}\mathbf{P}^{\mathrm{f}}_{N,\omega}}{\mathrm{d}\mathbf{P}}(S) = \frac{1}{Z^{\mathrm{f}}_{N,\omega}} \exp\left(\sum_{n=1}^{N} (\beta\omega_n + h)\delta_n \right), \tag{3.8}$$

where $\delta_n = \mathbf{1}_{S_n=0}$. One easily works out that S, under $P_{N,\omega}$, is still a Markov chain, but a inhomogeneous one, and for $n < N$ and we have the following formula:

$$\mathbf{P}^{\mathrm{f}}_{N,\omega}\left(S_{n+1} = m+a \middle| S_n = m\right) = e^{(\beta\omega_{n+1}+h)\mathbf{1}_{m+a=0}}\mathbf{P}(S_1 = a)\frac{\mathscr{Z}_{n+1,N,\omega}(m+a)}{\mathscr{Z}_{n,N,\omega}(m)}, \tag{3.9}$$

where

$$\mathscr{Z}_{n,N,\omega}(m) := \mathbf{E}\left[\exp\left(\sum_{j=n+1}^{N} (\beta\omega_j + h)\delta_j \right) \middle| S_n = m \right], \tag{3.10}$$

and we recall that $\mathbf{P}(S_1 = a) = (p/2)\mathbf{1}_{|a|=1} + q\mathbf{1}_{a=0}$. Therefore the transition probabilities of this Markov process depend *on (all, when $N \to \infty$) the future charges*.

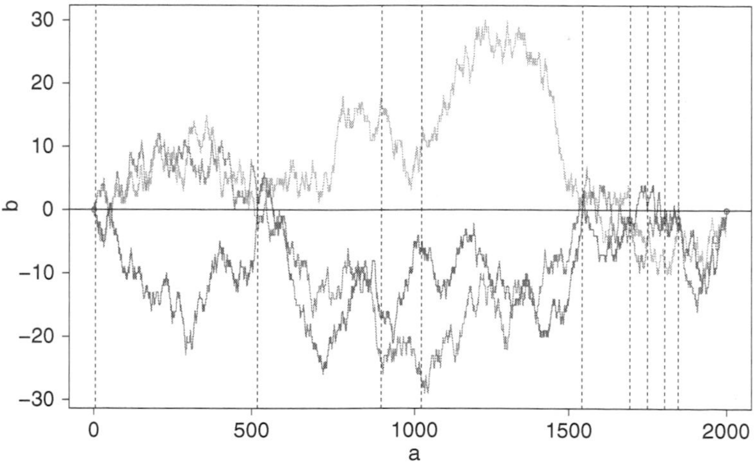

Fig. 3.1 This figure is meant to give an idea of the pinning effect that even typically repulsive charges may have. For the three trajectories we have chosen $\beta = 4$, $h = -5$ and $N = 2{,}000$ (the underlying walk has $p = q = 1/2$ and the charges are standard Gaussian). There are about 200 sites n with $\beta\omega_n + h > 0$ and nine sites, marked by vertical dashed lines, in which $\beta\omega_n + h > 5$

3.2 Super-Additivity, Free Energy, and Localization

Let us take a closer look at the free energy and let us point out from the start the self-averaging property.

Proposition 3.2. $\mathbb{P}(\mathrm{d}\omega)$*-a.s. and in* L^1 *we have that*

$$\lim_{N\to\infty}\frac{1}{N}\log Z_{N,\omega,\beta,h} = \mathrm{F}(\beta,h) \quad \textit{for every } \beta \textit{ and } h\,. \tag{3.11}$$

The proof is postponed to the next section. Here we rather point out a number of rather straightforward properties of the free energy:

1. $(\beta,h)\mapsto \mathrm{F}(\beta,h)$ is a convex function (since it is the limit of a sequence of convex functions).
2. $h\mapsto \mathrm{F}(\beta,h)$ is non-decreasing for every β, since $h\mapsto \mathrm{F}_N(\beta,h)$ clearly is.
3. Also $\beta\mapsto \mathrm{F}(\beta,h)$ is non-decreasing in $\beta(\geq 0)$: in fact $\partial_\beta \mathrm{F}_N(\beta,h)$ is equal to $(1/N)\mathbb{E}\mathbf{E}_{N,\omega}[\sum_{n=1}^N \omega_n\delta_n]$, so that $\partial_\beta \mathrm{F}_N(0,h)=0$ because the charges are centered and $\mathbf{E}_{N,\omega}$ does not depend on ω if $\beta=0$. Therefore, by convexity, $\beta\mapsto \mathrm{F}_N(\beta,h)$ is non decreasing on the positive semi-axis and the claim follows.
4. (*The annealed bound*). By Jensen inequality, the Fubini–Tonelli Theorem and by the IID character of ω we see that

$$\begin{aligned}\mathrm{F}_N(\beta,h) &\leq \frac{1}{N}\log\mathbf{E}\mathbb{E}\left[\exp\left(\sum_{n=1}^N(\beta\omega_n+h)\delta_n\right)\delta_N\right]\\ &= \frac{1}{N}\log\mathbf{E}\left[\exp\left(\sum_{n=1}^N(\log\mathrm{M}(\beta)+h)\delta_n\right)\delta_N\right] = \mathrm{F}_N(0,h+\log\mathrm{M}(\beta))\,,\end{aligned} \tag{3.12}$$

that, combined with the monotonicity of point (3) sandwiches the quenched free energy between two explicit quantities

$$\mathrm{F}(0,h)\leq \mathrm{F}(\beta,h)\leq \mathrm{F}(0,h+\log\mathrm{M}(\beta))\,. \tag{3.13}$$

The last formula is telling us in particular that $\mathrm{F}(\beta,h)=0$ if $h<h_c(0)-\log\mathrm{M}(\beta)$ ($h_c(0)$ is of course well known: it is just the critical value for the homogeneous system, that is $-\log\mathbf{P}(\tau_1<\infty)$) and that $\mathrm{F}(\beta,h)>0$ if $h>h_c(0)$. Therefore the transition between *localized and delocalized regime*, defined in strict analogy with the non disordered case (see Remark 3.3 and Fig. 3.2 below), lies in $[h_c(0)-\log\mathrm{M}(\beta),h_c(0)]$: the monotonicity of $\mathrm{F}(\beta,\cdot)$ directly shows that there exists a critical (quenched) value $h_c(\beta)$

$$h_c(\beta):=\inf\{h:\mathrm{F}(\beta,h)>0\} = \max\{h:\mathrm{F}(\beta,h)=0\}\,, \tag{3.14}$$

and we define

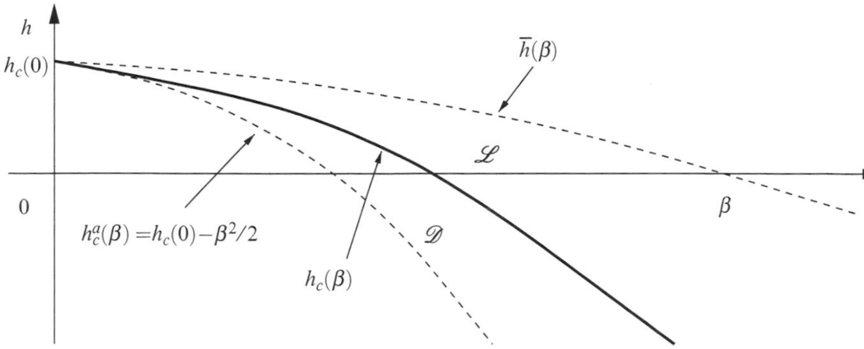

Fig. 3.2 The critical curve $\beta \mapsto h_c(\beta)$ that separates $\mathscr{D}$ and $\mathscr{L}$ is concave decreasing. This follows from the fact that $\mathscr{D}$ is a convex set and from the explicit bounds we have on the critical curve. We are not going to give an explicit expression for $\overline{h}(\cdot)$, we content ourselves with the fact that $\overline{h}(\beta) > h_c(0)$ for every $\beta > 0$, that is disorder may induce localization and it never suppresses it. The lower bound comes from the standard annealing procedure. Note that the annealed partition function $\mathbb{E}Z_{N,\omega}$ is just the homogeneous partition function with pinning potential $h + \log \mathrm{M}(\beta)$, so that we call $h_c^a(\beta)$ annealed critical point

$$\mathscr{L} := \{(\beta,h) : h > h_c(\beta)\} \quad \text{and} \quad \mathscr{D} := \{(\beta,h) : h \le h_c(\beta)\} . \tag{3.15}$$

Note that (3.13) directly yields

$$h_c^a(\beta) := h_c(0) - \log \mathrm{M}(\beta) \le h_c(\beta) \le h_c(0) . \tag{3.16}$$

Moreover property (1) guarantees that $\{(\beta,h) : \textsc{f}(\beta,h) \le 0\}$ is a convex set and this, coupled with (3.16), tells us that $h_c(\cdot)$ is concave.

Remark 3.3. We passed rather quickly over the fundamental fact that $\textsc{f}(\beta,h) \ge 0$ and that strictly positive (resp. vanishing) free energy *means* localization (resp. delocalization). A first justification comes by taking derivatives of the free energy with respect to h and by using convexity, like in (2.18). An analysis of the properties of the process is going to be taken up in some detail in Chap. 8. Here we want to add a more direct proof of $\textsc{f}(\beta,h) \ge 0$, with an explicit bound: note that

$$Z_{N,\omega} \ge Z_{N,\omega}(\tau_1 = N) = \exp(\beta\omega_N + h)\mathbf{P}(\tau_1 = N) , \tag{3.17}$$

immediately yields

$$\textsc{f}_N(\beta,h) \ge \frac{1}{N} \log \mathbf{P}(\tau_1 = N) + \frac{h}{N} , \tag{3.18}$$

which tells us that there exists $c = c(h, K(\cdot)) > 0$ such that for every $N \in \mathbb{N}$ we have

$$\textsc{f}_N(\beta,h) \ge -c \frac{\log(N+1)}{N} , \tag{3.19}$$

which is sensibly stronger than $\textsc{f}(\beta,h) \ge 0$.

The inequalities in (3.13) and (3.16) call for the questions: can they be made strict? Let us point out that the upper (respectively lower) bound in (3.13) correspond to the lower (respectively upper) bound in (3.16). So, in a sense, the problem splits:

- It has been shown in great generality that $h_c(\beta) < h_c(0)$ as soon as $\beta > 0$ [1] (see [6, Chap. 5, Sect. 2] for a more direct approach that yields the quantitative bound $h_c(0) - h_c(\beta) \geq c\beta^2$ for $\beta \in (0,1]$ and $c = c(K(\cdot), \mathbb{P}) > 0$ for a class of models that is less general than the one in [1], but it strictly includes all the cases that we treat in these notes). Since showing such a result requires showing that for every $\beta > 0$ there exists $h < h_c(0)$ such that $\textsc{f}(\beta, h) > 0$, we see that the left-most inequality in (3.13) can be made strict in full generality, as soon as $\beta > 0$.
- In general the lower bound in (3.16) cannot be replaced by a strict inequality and this is at the heart of the topics of these notes. We will see in fact that for $\alpha < 1/2$ we have $h_c(\beta) = h_c(0) - \log \mathrm{M}(\beta)$, at least if β is not too large. Otherwise stated, quenched and annealed critical points coincide. We will also show that this is not the case for $\alpha \geq 1/2$, and this directly implies that the upper bound in (3.13) is strict for $\alpha \geq 1/2$. Such an upper bound happens to be strict also for $\alpha < 1/2$, even if, in this case, annealed and quenched critical points coincide: we will come back to this in the next chapters.

Instead of summing up all the previous arguments, results and anticipations with a statement, we draw a figure (Fig. 3.2).

3.2.1 Two Important Remarks

3.2.1.1 A Crucial Corollary to Super-Additivity

There is still a very important observation, that is a direct corollary of the super-additivity property of $\{\mathbb{E} \log Z_{N,\omega}\}_N$ and therefore that $\textsc{f}(\beta, h) = \sup_N \frac{1}{N} \mathbb{E} \log Z_{N,\omega}$, that is (3.7):

Proposition 3.4. *We have* $(\beta, h) \in \mathscr{L}$ *if and only if there exists* $N \in \mathbb{N}$ *such that* $\mathbb{E} \log Z_{N,\omega} > 0$.

We can restate this proposition in a way that should make its relevance clear: if the systems is localized, this can be read off in finite volume.

3.2.1.2 Why Disorder Localizes?

We have just seen that disorder favors localization: $\textsc{f}(0, h) \leq \textsc{f}(\beta, h)$ by convexity and then we have referred to the literature for the strict inequality. It is however useful, in order to get a better grasp of the matter, to explain for example why no matter how negative h is and no matter how close to one $K(\infty)$ is, the system is localized if β is sufficiently large.

For this we can argue by introducing the important idea of *strategy* and of *entropy-energy competition*. Let us choose the strategy of selecting only the trajectories that visit only and all the non-negative charges: this cuts down the entropy of the system, but the energy gain can be made arbitrarily large by playing on β and overcome the entropy loss. Let us see it in detail: we can associate to every charge sequence ω the IID sequence of (translated) geometric random variables $\{X_j\}_{j\in\mathbb{N}}$ with success event $\omega_n > 0$, that is $X_1 = \inf\{n : \omega_n > 0\}$ and so on. Let us set $\iota_N := \sup\{n : \sum_{j=1}^n X_j \le N\}$, with $\sup\emptyset = 0$. We choose to deal with the free partition function because the proof turns out to be more compact (the fact, already announced, that free and constrained systems have the same free energy is proven in the next section):

$$Z_{N,\omega}^{\mathrm{f}} \ge \exp\left(\sum_{n=1}^N (\beta\omega_n + h)\mathbf{1}_{\omega_n\ge 0}\right)\left(\prod_{i=1}^{\iota_N} K(X_i)\right)\overline{K}\left(N - \sum_{i=1}^{\iota_N} X_i\right), \tag{3.20}$$

where $\prod_{i=1}^{\iota_N}\cdots = 1$ if $\iota_N = 0$. Therefore when $\iota_N > 0$

$$\frac{1}{N}\log Z_{N,\omega}^{\mathrm{f}} \ge \frac{1}{N}\sum_{n=1}^N (\beta\omega_n + h)\mathbf{1}_{\omega_n>0} + \frac{\iota_N}{N}\frac{1}{\iota_N}\sum_{i=1}^{\iota_N}\log K(X_i) - c\frac{\log N}{N}, \tag{3.21}$$

with $c > 0$. By the Law of Large Numbers $\lim_N \iota_N/N = \mathbb{P}(\omega_1 > 0) =: p$ almost surely and, again by the Law of Large Numbers, one easily goes from (3.21) to

$$\mathrm{F}(\beta,h) \ge \beta\,\mathbb{E}[\omega_1;\omega_1>0] + p\left(h + \sum_{n\in\mathbb{N}} p^{n-1}(1-p)\log K(n)\right), \tag{3.22}$$

and we see that for every choice of the charge distribution, every $K(\cdot)$ and every h we find an explicit β_0 guaranteeing that $\mathrm{F}(\beta,h) > 0$ as soon as $\beta > \beta_0$.

3.3 Self-Averaging Property, Effect of Boundary Condition

This section is devoted to the proof of Proposition 3.2 and to showing that the *pinned down* boundary condition does not affect the value of the free energy.

3.3.1 Proof of Proposition 3.2

Proposition 3.2 is a direct consequence of (3.5) and the Sub-additive (or Super-additive) Ergodic Theorem (see e.g. [8]). Actually, ergodic super-additivity yields the result well beyond IID charges. Another way to dispose quickly with such a proof would be to use concentration inequalities (at the expense however of imposing some conditions on the law of the charges and with the advantage of obtaining explicit bounds, see Chap. 8).

We take instead a very basic hand-on approach and we will just rely on Kolmogorov Law of Large Numbers. Note in fact that, by (3.5), for every $L \in \mathbb{N}$ and for $N/L \in \mathbb{N}$ we have

$$\frac{1}{N} \log Z_{N,\omega} \geq \frac{L}{N} \sum_{j=0}^{(N/L)-1} \frac{1}{L} \log Z_{L,\theta^{jL}\omega}, \tag{3.23}$$

and the right-hand side converges to $\mathrm{F}_L(\beta,h)$ $\mathbb{P}(\mathrm{d}\omega)$-a.s. and in L^1. Therefore it suffices to establish a sub-additive analog to the super-additive ergodic inequality (3.5). The argument argument goes as follows: for $M \in \{1,2,\ldots,N-1\}$ we can write

$$Z_{N,\omega} = Z_{M,\omega} Z_{N-M,\theta^M\omega} + \sum_{j=0}^{M-1} \sum_{k=M+1}^{N} Z_{j,\omega} K(k-j) \exp(\beta\omega_k + h) Z_{N-k,\theta^k\omega}, \tag{3.24}$$

and then we observe that there exists $C = C(K(\cdot)) > 0$ such that for every j, k, and M as in (3.24)

$$\frac{K(k-j)}{K(M-j)K(k-M)} \leq C((k-M) \wedge (M-j))^{1+\alpha} \leq C((N-M) \wedge M)^{1+\alpha}. \tag{3.25}$$

By inserting this bound into (3.24) and by performing some elementary estimates we obtain

$$\begin{aligned} Z_{N,\omega} &\leq Z_{M,\omega} Z_{N-M,\theta^M\omega} + Ce^{\beta|\omega_M|+|h|}((N-M) \wedge M)^{1+\alpha} \\ &\quad \times \sum_{j=0}^{M-1} \sum_{k=M+1}^{N} Z_{j,\omega} K(M-j) e^{\beta\omega_M + h} K(k-M) e^{\beta\omega_{N-k}+h} Z_{N-k,\theta^k\omega} \\ &\leq Z_{M,\omega} Z_{N-M,\theta^M\omega} + Ce^{\beta|\omega_M|+|h|}((N-M) \wedge M)^{1+\alpha} Z_{M,\omega} Z_{N-M,\theta^M\omega}, \end{aligned} \tag{3.26}$$

so that we have proven that for every β, every h and every $K(\cdot)$ there exists a constant $c > 0$ such that

$$\log Z_{N,\omega} \leq \log Z_{M,\omega} + \log Z_{N-M,\theta^M\omega} + c + \beta|\omega_M| + (1+\alpha)\log((N-M) \wedge M). \tag{3.27}$$

Now we choose again $L \in \mathbb{N}$ and for $N/L \in \mathbb{N}$ we have

$$\frac{1}{N} \log Z_{N,\omega} \leq \frac{L}{N} \sum_{j=0}^{(N/L)-1} \left(\frac{1}{L} \log Z_{L,\theta^{jL}\omega} + \frac{1}{L}(c + \beta|\omega_{jL}| + (1+\alpha)\log L) \right). \tag{3.28}$$

The Law of Large Numbers tells us that the right-hand side converges almost surely and in L^1 as $N \to \infty$ to $\mathrm{F}_L(\beta,h) + (c+\beta\mathbb{E}|\omega_1| + (1+\alpha)\log L)/L$.

Let us sum-up what we have done: for every chosen L we have sandwiched $(1/N)\log Z_{N,\omega}$ between two random variables that converge (a.s. and in L^1) to non random limits whose difference is $o(1)$ for L large: this would suffice, except that, to be precise, we haven't proven this for every N, but only for $N/L \in \mathbb{N}$. To extend the result to every N we proceed by a rough *cut and paste* procedure in the spirit of the one we have just performed to obtain that there exists $c_1 = c_1(K(\cdot)) \in (0,1)$ such that

$$\frac{c_1}{L^{1+\alpha}} \exp\left(-\sum_{n\in I_j}(\beta|\omega_n| + |h|)\right) \le \frac{Z_{jL+m,\omega}}{Z_{jL,\omega}} \le \frac{L^{1+\alpha}}{c_1}\exp\left(\sum_{n\in I_j}(\beta|\omega_n|+|h|)\right), \tag{3.29}$$

where $I_j := \{jL+1,\ldots,(j+1)L\}$ and (3.29) holds for every $j \in \mathbb{N}\cup\{0\}$, $m \in \{0,1,\ldots,L-1\}$ and every ω. By applying once again the Law of Large Numbers (or, rather, the equivalent statement that for L^1 IID variables we have $\lim_{n\to\infty} X_n/n = 0$ a.s. and, of course, in L^1) one easily sees that

$$\limsup_{N\to\infty}\left|\frac{1}{N}\log\frac{Z_{N,\omega}}{Z_{L\lfloor N/L\rfloor,\omega}}\right| \le \frac{1}{L}\left(\log(L^{1+\alpha}/c_1) + \beta\mathbb{E}|\omega_1| + |h|\right), \tag{3.30}$$

almost surely and of course the same inequality holds also if we replace $\limsup|\ldots|$ by $\limsup\mathbb{E}|\ldots|$. The proof of Proposition 3.2 is therefore complete. □

3.3.2 *Free and Constrained Models*

The following model

$$\frac{\mathrm{d}\mathbf{P}^{\mathrm{f}}_{N,\omega}}{\mathrm{d}\mathbf{P}}(\tau) = \frac{1}{Z^{\mathrm{f}}_{N,\omega}}\exp\left(\sum_{n=1}^{N}(\beta\omega_n + h)\delta_n\right), \tag{3.31}$$

is obviously a close companion of (3.1). Let us *quantify* the difference between these two models by looking at the Laplace asymptotic behavior of the partition function, that in the free case is

$$Z^{\mathrm{f}}_{N,\omega} = \mathbf{E}\left[\exp\left(\sum_{n=1}^{N}(\beta\omega_n + h)\delta_n\right)\right], \tag{3.32}$$

with $Z^{\mathrm{f}}_{0,\omega}$ to be read as 1. A direct consequence of Proposition 3.2 and of Lemma 3.5 below is that

$$\lim_{N\to\infty}\frac{1}{N}\log Z^{\mathrm{f}}_{N,\omega} = \mathrm{F}(\beta,h). \tag{3.33}$$

Lemma 3.5. *For every $K(\cdot)$ as in* (2.30) *there exists $C = C(K(\cdot))$ such that for every $N = 0, 1, 2, \dots$*

$$Z_{N,\omega} \le Z^{\mathtt{f}}_{N,\omega} \le Z_{N,\omega}\left(1 + CN^q \exp(-\beta\omega_N - h)\right), \tag{3.34}$$

where $q = 1$ if $\sum_n K(n) = 1$ and $q = 1 + \alpha$ if $\sum_n K(n) < 1$.

Proof. For $N = 0$ the three terms are equal to one. Let us therefore assume $N \ge 1$. We write

$$Z^{\mathtt{f}}_{N,\omega} = Z_{N,\omega} + \sum_{n=0}^{N-1} Z_{n,\omega}\left(\overline{K}(N-n) + K(\infty)\right), \tag{3.35}$$

from which we see that the lower bound holds, but also that

$$\begin{aligned} Z^{\mathtt{f}}_{N,\omega} &\le Z_{N,\omega} + \max_{j=1,\dots,N}\left(\frac{\overline{K}(j) + K(\infty)}{K(j)}\right)\sum_{n=0}^{N-1} Z_{n,\omega}K(N-n) \\ &= Z_{N,\omega} + \max_{j=1,\dots,N}\left(\frac{\overline{K}(j) + K(\infty)}{K(j)}\right) Z_{N,\omega}\exp(-\beta\omega_N - h). \end{aligned} \tag{3.36}$$

Since if $K(\infty) = 0$ we have

$$\frac{\overline{K}(j)}{K(j)} \stackrel{j\to\infty}{\sim} \frac{j}{\alpha}, \tag{3.37}$$

and if $K(\infty) > 0$

$$\frac{\overline{K}(j) + K(\infty)}{K(j)} \stackrel{j\in\mathbb{N}}{\le} \frac{1}{K(j)} \stackrel{j\to\infty}{\sim} \frac{j^{1+\alpha}}{c_K}, \tag{3.38}$$

we are done. □

3.4 A Look at the Literature and, Once Again, Correlation Length Issues

There is an extended physical literature on disordered pinning models: here we content ourselves with citing [5] and [4] that have been very influential.

The mathematical techniques used in this chapter are rather standard in equilibrium statistical mechanics of disordered systems (see e.g. [2, 3]), but, at the same time, rather model dependent, with plenty of renewal process estimates that are specific to pinning models [6, 7]. Super/sub-additive arguments are absolutely crucial and, even if no direct use of Kingman Ergodic Sub-additive Theorem [8] has been made, we fully acknowledge the pivotal role of Kingman's ideas.

In Sect. 2.4 we have introduced and discussed the notion of correlation length for the homogeneous model and we have set forth the fact that the natural correlation length is simply $1/\mathrm{F}(h)$ (with the disordered model notation we would write

$1/\text{F}(0,h)$). This settles the issue for the correlation length of the annealed model, since it is simply a homogeneous model. But what about the quenched model? This is an important point an will be taken up fully only in Chap. 8. For now we signal the detailed presentation in [9] and observe that

- $1/\text{F}(\beta,h)$ is *one of* the natural correlation lengths of disordered models, but not the only one: there are in fact subtle and very interesting issues connected to choosing to look at quenched quantities or to quenched quantities averaged over the disorder (this is at first surprising, because the free energy is self-averaging, see however Chap. 8).
- In the mathematical arguments that we present in the sequel the correlation length of the annealed system plays a more central role than its quenched companions, ultimately because our main results are establishing that the quenched system is either close or not close to the annealed one.

Figure 3.3 proposes a quantitative view on self-averaging and correlation length.

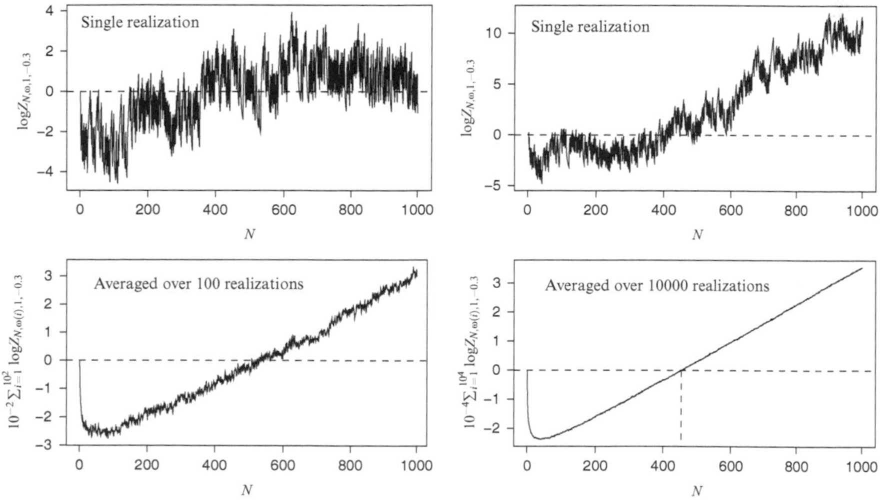

Fig. 3.3 The first two figures (on *top*) represent the graph of $\log Z_{N,\omega}$, for N up to 1,000, with two different realizations of the charge sequence ω. In both cases, and in the following two as well, $\beta = 1$, $h = -0.3$ and ω_1 takes only the values ± 1, with equal probability. The underlying process is the renewal associated to a random walk that stays at zero with probability $1/2$ or jumps of ± 1, with probability $1/4$ each. The profile of $N \mapsto \mathbb{E}\log Z_{N,\omega}$ starts appearing when one averages over several (independent) realizations ($\omega(i)$, $i = 1,2,\dots$, are the realizations). As a matter of fact 100 realizations give a result that is rather satisfactory and 10,000 realization give a very satisfactory result (according to the fourth graph $\mathbb{E}\log Z_{N,\omega} > 0$ starting from $N =$ 454). Note that a rough estimate of the slope suggests $\text{F}(1,-0.3) \approx 6 \times 10^{-3}$, which yields a correlation length ($1/\text{F}(\beta,h)$, in analogy with Sect. 2.4) close to 200, which is compatible with the *other* correlation length 454. Precise quantitative estimates on this averaging procedure can be obtained by concentration inequalities (see Chap. 8)

References

1. K.S. Alexander, V. Sidoravicius, Pinning of polymers and interfaces by random potentials. Ann. Appl. Probab. **16**, 636–669 (2006)
2. E. Bolthausen, A.-S. Sznitman, *Ten Lectures on Random Media*. DMV Seminar, vol. 32 (Birkhäuser, Switzerland, 2002)
3. A. Bovier, *Statistical Mechanics of Disordered Systems. A Mathematical Perspective*. Cambridge Series in Statistical and Probabilistic Mathematics (Cambridge University Press, Cambridge, 2006)
4. B. Derrida, V. Hakim, J. Vannimenus, Effect of disorder on two-dimensional wetting. J. Stat. Phys. **66**, 1189–1213 (1992)
5. G. Forgacs, J.M. Luck, Th. M. Nieuwenhuizen, H. Orland, Wetting of a disordered substrate: exact critical behavior in two dimensions. Phys. Rev. Lett. **57**, 2184–2187 (1986)
6. G. Giacomin, *Random Polymer Models* (Imperial College Press, London, 2007)
7. F. den Hollander, *Random polymers*, in Lectures from the 37th Probability Summer School held in Saint-Flour, 2007, Lecture Notes in Mathematics, vol. 1974 (Springer, Berlin, 2009)
8. J.F.C. Kingman, Subadditive ergodic theory. Ann. Probab. **1**, 882–909 (1973)
9. F.L. Toninelli, *Localization transition in disordered pinning models. Effect of randomness on the critical properties*, in Methods of Contemporary Mathematical Statistical Physics, Lecture Notes in Mathematics, vol. 1970 (2009), pp. 129–176

Chapter 4
Irrelevant Disorder Estimates

Abstract We introduce the Harris criterion and we provide two heuristic arguments in favor of this criterion. In particular we introduce the notion of relevant and irrelevant disorder. We then prove that disorder is irrelevant when the inter-arrival exponent α is smaller than $1/2$ and β is not too large.

4.1 Disorder and Critical Behavior: What to Expect?

The questions we want to address are:

1. Can one compute $h_c(\beta)$ for $\beta > 0$?
2. Can one determine the critical behavior of the free energy at criticality, namely the way $\mathrm{F}(\beta, h)$ vanishes as $h \searrow h_c(\beta)$?

Such questions find partial answers in the physical literature and in spite of the lack of mathematical rigor and of the, possibly more serious, problem that at times one finds multiple (non-coinciding) answers to a given question, this is definitely a starting point. Particularly interesting for us is a claim, going under the name of *Harris criterion*, that can be resumed by the rather suggestive statement that whether or not introducing a small (or moderate) amount of disorder changes the critical behavior of a system should be read off the critical behavior of the *pure*, i.e. non-disordered, system. This criterion has been proposed by Harris in the context of diluted Ising model [8] and can be resumed by saying that a small amount of disorder does not change the critical behavior of the model if the *specific heat exponent is negative*, while it is expected to change it if it is positive. We will go into the detail of such a statement for the models we consider (and then for more general models too), but let us keep vague still for a while and say that the idea behind Harris' approach is *renormalization*. The renormalization procedure is not something uniquely defined and for our discussion it is just a *coarse graining* transformation meant to map a given model to a rougher one. Successive applications of this transformation are supposed to drive the system to a fixed point (in the *space of models*) and such a

G. Giacomin, *Disorder and Critical Phenomena Through Basic Probability Models*, Lecture Notes in Mathematics 2025, DOI 10.1007/978-3-642-21156-0_4,
© Springer-Verlag Berlin Heidelberg 2011

fixed point is *trivial* unless one sits at the critical point. In the disordered case one introduces also the further idea that the disorder may, or may not, be suppressed by the successive renormalization transformations. This brings into the game the notion of irrelevant (respectively, relevant) disorder, that in the end boils down to a critical behavior which is not affected (respectively, is affected) by the presence of disorder. In general, it is very difficult to make concrete such a scheme and physicists often resort to ingenious arguments to capture the flavor of what the renormalization flow leads. For pinning models this procedure has been taken up by several authors, following the two seminal works by Forgacs, Luck, Nieuwenhuizen and Orland [6] and by Derrida, Hakim and Vannimenus [5]. What I will do next is to present two arguments, that correspond or that are inspired by the two papers I just mentioned. Let us anticipate what these two arguments suggest in the end:

1. Disorder is irrelevant if $\alpha < 1/2$ and $\beta \leq \beta_0$, for some $\beta_0 > 0$, in the sense not only that the critical behavior is not modified by the disorder, but also that $h_c(\beta)$ coincides with the critical point of the annealed system $h_c^a(\beta) = h_c(0) - \log \mathrm{M}(\beta)$ (recall that the free energy of the annealed system is $\mathrm{F}(0, h + \log \mathrm{M}(\beta))$). Since the critical behavior of the non-disordered model coincides with the critical behavior of the annealed model, in the sequel we will use *pure model* as a synonym of *annealed model*.
2. In a somewhat weaker way they suggest also that disorder is relevant if $\alpha > 1/2$, in the sense that quenched and annealed critical points differ and that *one does not really see a reason* why the critical behavior should coincide (however, these procedures give no hint as to what the new critical behavior should be).
3. The $\alpha = 1/2$ is somewhat *undecidable* at the level of Harris' approach and it is dubbed *marginal*. Still, the question of whether or not the disorder changes the critical behavior is there and the two approaches in this case suggest a different answer (while they yield the same prediction for $\alpha \neq 1/2$).

Without loss of generality we assume $\sum_n K(n) = 1$, hence $h_c(0) = 0$ and $h_c^a(\beta) = -\log \mathrm{M}(\beta)$.

4.1.1 First Approach: An Expansion in Powers of β^2

We fix a value of $\beta > 0$ and choose $h - h_c^a(\beta) = \Delta \geq 0$ and we observe that

$$\mathbb{E}\left[Z_{N,\omega,\beta,h}^{\mathbf{f}}\right] = \mathbf{E}\left[\exp\left(\Delta |\tau \cap (0,N]|\right)\right] = Z_{N,\Delta}^{\mathbf{f}}. \tag{4.1}$$

We introduce the centered random variables $\zeta_n = \exp(\beta\omega_n - \log \mathrm{M}(\beta)) - 1$ and observe that

$$
\begin{aligned}
&\mathbb{E}\log \frac{Z^{\mathtt{f}}_{N,\omega}}{\mathbb{E}Z^{\mathtt{f}}_{N,\omega}} \\
&= \mathbb{E}\log \mathbf{E}^{\mathtt{f}}_{N,\Delta}\left[\exp\left(\sum_{n=1}^{N}(\beta\omega_n - \log \mathrm{M}(\beta))\delta_n\right)\right] = \mathbb{E}\log \mathbf{E}^{\mathtt{f}}_{N,\Delta}\left[\prod_{n=1}^{N}(1+\zeta_n\delta_n)\right] \\
&= \mathbb{E}\log\left(1+\sum_n \zeta_n \mathbf{P}^{\mathtt{f}}_{N,\Delta}(n\in\tau) + \sum_{n_1<n_2}\zeta_{n_1}\zeta_{n_2}\mathbf{P}^{\mathtt{f}}_{N,\Delta}(\{n_1,n_2\}\subset\tau)+\ldots\right). \qquad (4.2)
\end{aligned}
$$

Let us now expand the logarithm and use the fact that the ζ_n's are IID, centered and of variance $\sigma^2(\beta) := \mathrm{M}(2\beta)/\mathrm{M}(\beta)^2 - 1$ (may be useful to note that $\sigma^2(\cdot)$ is increasing and that $\sigma^2(\beta) \sim \beta^2$ as $\beta \searrow 0$) to see that

$$
\mathbb{E}\log \frac{Z^{\mathtt{f}}_{N,\omega}}{\mathbb{E}Z^{\mathtt{f}}_{N,\omega}} = -\frac{1}{2}\sigma^2(\beta)\sum_{n=1}^{N}\mathbf{P}^{\mathtt{f}}_{N,\omega}(n\in\tau)^2 + \ldots \qquad (4.3)
$$

Of course for $\Delta > 0$ and as long as n and $N-n$ are large, $\mathbf{P}^{\mathtt{f}}_{N,\Delta}(n\in\tau)$ is close to $\partial_\Delta \mathrm{F}(0,\Delta)$ (this can be proven by using Renewal Theorem and by applying the arguments in the proof of Proposition 2.5) so that from (4.2) we extract

$$
\mathrm{F}(\beta, h_c^a(\beta)+\Delta) = \mathrm{F}(0,\Delta) - \frac{1}{2}\sigma^2(\beta)\left(\partial_\Delta \mathrm{F}(0,\Delta)\right)^2 + \ldots \qquad (4.4)
$$

This is just a formal expansion (which can be continued) and we do not want to try to justify it. Rather we remark that this computation is compatible with $h_c(\beta) = h_c^a(\beta)$ if $(\partial_\Delta \mathrm{F}(0,\Delta))^2$ is of the order (or possibly much smaller) than $\mathrm{F}(0,\Delta)$, when $\Delta \searrow 0$. Since for $\alpha \in (0,1)$ we have $\mathrm{F}(0,\Delta) \sim (const.)\Delta^{1/\alpha}$ and $\partial_\Delta \mathrm{F}(0,\Delta) \sim (const.)\Delta^{-1+1/\alpha}$, this argument suggests that $h_c(\beta) = h_c^a(\beta)$ and that the quenched critical exponent coincides with the annealed one if $\Delta^{2(-1+1/\alpha)}$ is much smaller than (or, possibly, of the same order of) $\Delta^{1/\alpha}$, that is if $\alpha < 1/2$ (or $\alpha \le 1/2$).

In Harris' language, this argument therefore suggests that disorder is *irrelevant* for $\alpha < 1/2$ (the situation for $\alpha = 1/2$ is delicate and even at a heuristic level one should be careful: in [6] one can find an expansion in powers of $\sigma^2(\beta)$ to all orders and the claim that disorder is irrelevant also at $\alpha = 1/2$).

If $\alpha > 1/2$ the second term in the expansion is much larger than the first and if this is the trend (it is!) then something is going wrong. In this case we may argue that this is just due to the fact that $h_c(\beta) > h_c^a(\beta)$ and we are expanding around the *wrong* value of h. Pushing the argument farther we can observe that we do know that $\mathrm{F}(\beta, h_c(\beta)) = 0$ and therefore (4.4) suggests that for β small the shift of the quenched critical point $\Delta_c(\beta) := h_c(\beta) - h_c^a(\beta)$ can be found by equating the two terms in the rightmost side of (4.4): so $\Delta_c(\beta) \approx \beta^{2\alpha/(2\alpha-1)}$ for β small. This last argument might give the impression that we have pushed our luck too far: it turns out that it is not the case! See Chap. 6.

4.1.2 Second Approach: A 2-Replica Argument

Since we aim at deciding whether the annealed system is close to the quenched system we choose to sit at the annealed critical point [$h = -\log \mathrm{M}(\beta)$, i.e. $\Delta = 0$ with the convention set forth just before (4.1)] and study the variance of $Z^{\mathtt{f}}_{N,\omega}$. Divergence of the variance, as $N \to \infty$, would suggest that the disorder drives the quenched system away from the annealed one. On the other hand, if the variance is bounded in N we have a clear signal of the proximity of the quenched and annealed systems. Since at the annealed critical point we have $\mathbb{E}Z^{\mathtt{f}}_{N,\omega} = 1$ and

$$\begin{aligned}\mathrm{var}_{\mathbb{P}}\left(Z^{\mathtt{f}}_{N,\omega}\right) &= \mathbb{E}\left[\left(Z^{\mathtt{f}}_{N,\omega}\right)^2 - 1\right] \\ &= \mathbb{E}\mathbf{E}^{\otimes 2}\left[\exp\left(\sum_n (\beta\omega_n - \log \mathrm{M}(\beta))(\mathbf{1}_{n\in\tau} + \mathbf{1}_{n\in\tau'})\right) - 1\right],\end{aligned} \tag{4.5}$$

with τ and τ' independent copies of the same renewal process (the two *replica*). Integrating out the ω variables we obtain

$$\mathrm{var}_{\mathbb{P}}\left(Z^{\mathtt{f}}_{N,\omega}\right) = \mathbf{E}^{\otimes 2}\left[\exp\left(\lambda(\beta)\sum_{n=1}^{N} \mathbf{1}_{n\in\tau\cap\tau'}\right) - 1\right], \quad \text{with } \lambda(\beta) := \log\frac{\mathrm{M}(2\beta)}{\mathrm{M}(\beta)^2}. \tag{4.6}$$

Note that $\lambda(\beta) = \log(1+\sigma^2(\beta))$ is increasing and $\lambda(\beta) \sim \beta^2$ for $\beta \searrow 0$. The expression in (4.6) can be evaluated in a sharp way because the random set $\tau \cap \tau'$ is still a renewal process and therefore the variance that we are evaluating is the partition function of a homogeneous pinning model (minus one). What do we know of the $\tau\cap\tau'$ renewal? In principle everything, in practice it is not straightforward to write the most basic quantity, that is $\mathbf{P}^{\otimes 2}((\tau\cap\tau')_1 = n)$. But what is easy to write is the renewal function: independence guarantees that

$$\mathbf{P}^{\otimes 2}\left(n \in \tau\cap\tau'\right) = \mathbf{P}(n\in\tau)^2, \tag{4.7}$$

and this is largely sufficient for our purposes. In fact

$$\mathbf{E}^{\otimes 2}\left[|\tau\cap\tau'| - 1\right] = \sum_{n\in\mathbb{N}} \mathbf{P}(n\in\tau)^2 =: \gamma_2 \in (0,\infty], \tag{4.8}$$

Note that by Theorem A.4

$$\gamma_2 < \infty \;\Leftrightarrow\; \sum_n \frac{1}{n^{2(1-\alpha)}} < \infty \;\Leftrightarrow\; \alpha < \frac{1}{2}, \tag{4.9}$$

so that, for $\alpha < 1/2$, (4.8) tells us that $\tau\cap\tau'$ is terminating and, therefore, $|\tau\cap\tau'|$ is a geometric random variable of mean $1+\gamma_2$ or, equivalently, of parameter (i.e. success probability) $p_2 = 1/(1+\gamma_2)$. So for $\alpha < 1/2$ we have

$$\begin{aligned}\lim_{N\to\infty} \mathrm{var}_{\mathbb{P}}\left(Z^{\mathtt{f}}_{N,\omega}\right) &= \mathbf{E}^{\otimes 2}\left[\exp\left(\lambda(\beta)|\tau\cap\tau'|\right)\right]-1 \\ &= \frac{p_2}{1-(1-p_2)\exp(\lambda(\beta))}-1 \overset{\beta\searrow 0}{\sim} \gamma_2\beta^2 .\end{aligned} \tag{4.10}$$

Note that the second equality in (4.10) holds only if the denominator is positive, otherwise the expression in (4.10) is equal to ∞. And the denominator is positive if

$$\beta < \beta_0 := \lambda^{-1}(\log((1+\gamma_2)/\gamma_2)), \tag{4.11}$$

where $\lambda^{-1}(\log((1+\gamma_2)/\gamma_2))$ should be read as ∞ if the image of $\lambda(\cdot)$ does not contain $\log((1+\gamma_2)/\gamma_2)$. Note that it is not difficult to exhibit cases in which $\beta_0=\infty$.

Remark 4.1. By (4.8) and (4.9), $\alpha \geq 1/2$ is necessary and sufficient for persistence. In this case one can easily compute [use (A.3) and (2.9)] the value of

$$\lim_{N\to\infty}\frac{1}{N}\log \mathbf{E}^{\otimes 2}\left[\exp\left(\lambda(\beta)\sum_{n=1}^N \mathbf{1}_{n\in\tau\cap\tau'}\right)\right] > 0. \tag{4.12}$$

Exponential growth is there also if $\alpha<1/2$ but $\beta>\beta_0$. If $\alpha<1/2$ and $\beta=\beta_0$ the variance tends to infinity when $N\to\infty$, but not exponentially.

In a nutshell: the variance of $Z^{\mathtt{f}}_{N,\omega}$, at the critical annealed point, stays bounded if and only if $\alpha<1/2$ and if β is smaller than the threshold β_0 that we have computed. While we are in the middle of a heuristic argument, the computation we have just performed is rigorous and will be needed later, so let us write a statement.

Lemma 4.2. *Choose $K(\cdot)$ such that $\sum_n K(n)=1$. We have that*

$$\sup_N \mathbb{E}\left[\left(Z^{\mathtt{f}}_{N,\omega,\beta,h^a_c(\beta)}\right)^2\right] < \infty, \tag{4.13}$$

if and only if β is chosen as in (4.11) *[$\lambda(\cdot)$ the increasing function given in* (4.6)*].*

The way we are going to use Lemma 4.2 is via the observation:

$$\sup_N \mathbb{E}\left[Z^{\mathtt{f}}_{N,\omega,\beta,h^a_c(\beta)};\, Z^{\mathtt{f}}_{N,\omega,\beta,h^a_c(\beta)}>K\right] \leq \frac{1}{K}\sup_N \mathbb{E}\left[\left(Z^{\mathtt{f}}_{N,\omega,\beta,h^a_c(\beta)}\right)^2\right] \overset{K\to\infty}{\longrightarrow} 0. \tag{4.14}$$

Remark 4.3. While we have presented the second approach in a different way, it may be viewed also as an expansion around the annealed system. Observe in fact that

$$\begin{aligned}&\mathbb{E}\log Z_{N,\omega} - \log \mathbb{E} Z_{N,\omega} \\ &\quad = \mathbb{E}\left[\log\left(1+\left(\frac{Z_{N,\omega}}{\mathbb{E}Z_{N,\omega}}-1\right)\right)\right] = -\frac{1}{2}\mathrm{var}_{\mathbb{P}}\left(\frac{Z_{N,\omega}}{\mathbb{E}Z_{N,\omega}}\right)+\ldots,\end{aligned} \tag{4.15}$$

where we have of course used $\log(1+x)=x-x^2/2+\ldots$.

Remark 4.4. An attentive scrutiny of the two approaches we have presented reveals that they do not reach the same conclusion if $\alpha = 1/2$. This can be better appreciated in a larger class of models, namely if we consider for example $K(n) \sim (\log n)^a / n^{3/2}$, $a \in \mathbb{R}$. In fact $a > 0$ does not imply that $\tau \cap \tau'$ is terminating (one needs $a > 1/2$), while it is sufficient to conclude that $\mathrm{F}(0,\delta)$ is much larger that $(\partial_\delta \mathrm{F}(0,\delta))^2$ for δ small. As a matter of fact, as already pointed out we are dealing with the marginal case in the renormalization group sense and this is a subtle, and model dependent, issue. We should stress that the steps that we have just presented here are just a part of the arguments in [6] and [5], in particular [6] aims at an expansion to all orders and [5] contains an argument to capture the renormalization group flow for Δ close to 0. Both [6] and [5] consider only the case of $a = 0$ and their predictions differ: [6] stands for marginal irrelevance and [5] stands for marginal relevance.

4.2 Disorder is Irrelevant if $\alpha < 1/2$ (and if β is Not Too Large): A Proof

We take an approach due to Lacoin [9] that yields results that are weaker (see however Remark 4.8) than the ones of the original approaches [1, 10], but it is substantially simpler and it works for general disorder distribution. Our presentation differs from the one in [9] in the sense that we avoid using martingales and 0–1 law (see Remark 4.8).

Theorem 4.5. *If $\alpha \in (0,1/2)$ we can exhibit $\beta_0 \in (0,\infty]$ [see (4.11)] such that $h_c(\beta) = h_c^a(\beta)$ for every $\beta \in (0,\beta_0)$ and, for the same values of β, we have also*

$$\lim_{h \searrow h_c(\beta)} \frac{\log \mathrm{F}(\beta,h)}{\log (h - h_c(\beta))} = \frac{1}{\alpha}. \tag{4.16}$$

Let us start by proving two technical lemmas: the first is a basic probability result.

Lemma 4.6. *If $\{X_n\}_n$ is a sequence of random variables which is uniformly integrable (i.e. $\sup_n E[|X_n|; |X_n| > K]$ vanishes as K tends to ∞) and such that $EX_n = 1$ for every n, then for every $a \in (0,1)$ there exists $\delta > 0$ such that*

$$\inf_n P(X_n > a) > \delta. \tag{4.17}$$

Proof. Let us set $g(K) := \sup_n E[X_n; X_n > K]$. Uniform integrability directly implies that $\lim_{K\to\infty} g(K) = 0$. For $K \geq 1$ we have

$$\begin{aligned} 1 - g(K) &\leq E[X_n; X_n \leq a] + E[X_n; X_n \in (a,K]] \\ &\leq aP(X_n \leq a) + K(1 - P(X_n \leq a)), \end{aligned} \tag{4.18}$$

for every n, that is

$$\sup_n P(X_n \le a) \le \frac{K-1+g(K)}{K-a}. \tag{4.19}$$

The result is then achieved by choosing K large enough to have, for example $1-g(K) \ge (1+a)/2$. □

Lemma 4.6 implies a second result, which we now state for a general sequence of measurable events $A_N \in \mathscr{G}_\infty$ (that is an event that depends on all τ, cf. Sect. A.1.4 of Appendix A), but it will be applied to the case $A_N \in \mathscr{G}_N$ (that is an event that depends on $\tau \cap (0,N]$).

Lemma 4.7. *Let us fix $h = h_c^a(\beta)$, that is the subscript "N,ω" should be read as "$N,\omega,\beta,h_c^a(\beta)$" in the statement and in the proof. Let $\{A_N\}_N$ be a sequence such that $A_N \in \mathscr{G}_\infty$ and such that $\lim_n \mathbf{P}(A_N) = 0$. If $\left\{Z_{N,\omega}^{\mathbf{f}}\right\}_N$ is uniformly integrable, then*

$$\lim_{N\to\infty} \mathbf{P}_{N,\omega}^{\mathbf{f}}(A_N)\,\mathbf{1}_{Z_{N,\omega}^{\mathbf{f}}>1/2} = 0 \quad \textit{in } \mathbb{P}\textit{-probability}, \tag{4.20}$$

and there exists $\delta > 0$ such that

$$\inf_N \mathbb{P}\left(Z_{N,\omega}^{\mathbf{f}} > 1/2\right) > \delta\,. \tag{4.21}$$

The value of δ depends on the uniform integrability properties of $\{Z_{N,\omega}^{\mathbf{f}}\}_N$, that is, on the function $g(\cdot)$ used in the proof of Lemma 4.6: so, in the end, $\delta = \delta(\beta,\mathbb{P})$.

Proof. Since $\mathbb{E}Z_{N,\omega}^{\mathbf{f}} = 1$, cf. (4.1), and since we are assuming uniform integrability, (4.21) follows from Lemma 4.6. For what concerns (4.20) we use

$$\mathbb{E}\left[Z_{N,\omega}^{\mathbf{f}}\mathbf{P}_{N,\omega}^{\mathbf{f}}(A_N)\right] = \mathbb{E}\mathbf{E}\left[\exp\left(\sum_{n=1}^N (\beta\omega_n - \log \mathrm{M}(\beta))\delta_n\right)\mathbf{1}_{A_N}(\tau)\right] = \mathbf{P}(A_N)\,, \tag{4.22}$$

so that $Z_{N,\omega}^{\mathbf{f}}\mathbf{P}_{N,\omega}^{\mathbf{f}}(A_N)$ tends to zero, as N tends to infinity, in probability. By restricting the attention to the event $Z_{N,\omega}^{\mathbf{f}} > 1/2$ we obtain the claim. □

Proof of Theorem 4.5. Let us start by observing (rather, recalling) that both $h_c(\beta) \ge h_c^a(\beta)$ and that (4.16) with "=" replaced by "≥" and the limit by the inferior limit are immediate consequences of the annealed (upper) bound $\mathrm{F}(\beta,h) \le \mathrm{F}(0,h-h_c^a(\beta))$. So it suffices to deal with the lower bound. And it is worthwhile to stress by now that super-additivity properties, recall (3.7) and Proposition 3.4, reduce lower bounds on the free energy, obtained in the limit $N \to \infty$, to estimates for one (suitably chosen) value of N.

And let us start by observing that the uniform integrability of the sequence $\{Z_{N,\omega,\beta,h_c^a(\beta)}^{\mathbf{f}}\}_N$ is a direct consequence of Lemma 4.2. Now the point is to select the sequence of events A_N to which to apply Lemma 4.7. For this choose $\varepsilon \in (0,1/2)$ and introduce

$$A_N = A_N(\varepsilon) := \left\{|\tau \cap (0,N]| \le N^{\alpha(1-\varepsilon)}\right\}, \tag{4.23}$$

From Proposition A.6 we directly extract $\lim_N \mathbf{P}(A_N) = 0$. Moreover

$$\begin{aligned} Z^{\mathtt{f}}_{N,\omega,\beta,h^a_c(\beta)+h} &= Z^{\mathtt{f}}_{N,\omega,\beta,h^a_c(\beta)} \mathbf{E}^{\mathtt{f}}_{N,\omega,\beta,h^a_c(\beta)} \left[\exp\left(h|\tau \cap (0,N]|\right)\right] \\ &\geq Z^{\mathtt{f}}_{N,\omega,\beta,h^a_c(\beta)} \mathbf{P}^{\mathtt{f}}_{N,\omega,\beta,h^a_c(\beta)} \left(A_N^{\complement}\right) \exp\left(hN^{\alpha(1-\varepsilon)}\right). \end{aligned} \tag{4.24}$$

We now observe that $hN^{\alpha(1-\varepsilon)} \geq N^{\alpha\varepsilon}$ is equivalent to $h^{-1/((1-2\varepsilon)\alpha)} \leq N$ so that we make the choice

$$N := \left\lceil h^{-1/((1-2\varepsilon)\alpha)} \right\rceil, \tag{4.25}$$

and on the event $\{\omega : Z^{\mathtt{f}}_{N,\omega,\beta,h^a_c(\beta)} > 1/2\} \cap \{\omega : \mathbf{P}^{\mathtt{f}}_{N,\omega,\beta,h^a_c(\beta)}(A_N(\varepsilon)) \leq 1/2\}$ – note that, by Lemma 4.7, the probability of such an event is at least δ in the limit $N \to \infty$, so it is at least $\delta/2$ for h small – we have

$$Z^{\mathtt{f}}_{N,\omega,\beta,h^a_c(\beta)+h} \geq \frac{1}{2}\left(1 - \mathbf{P}^{\mathtt{f}}_{N,\omega,\beta,h^a_c(\beta)}\left(A_N(\varepsilon)\right)\right) \exp\left(N^{\alpha\varepsilon}\right) \geq \frac{1}{4} \exp\left(N^{\alpha\varepsilon}\right), \tag{4.26}$$

where in the second inequality we have applied again Lemma 4.7 with N sufficiently large (that is h sufficiently small) to guarantee that $\mathbf{P}^{\mathtt{f}}_{N,\omega,\beta,h^a_c(\beta)}(A_N(\varepsilon)) \leq 1/2$. We therefore see that

$$\mathbb{P}\left(Z^{\mathtt{f}}_{N,\omega,\beta,h^a_c(\beta)+h} \geq \frac{1}{4}\exp\left(N^{\alpha\varepsilon}\right)\right) \geq \frac{\delta}{2}. \tag{4.27}$$

We are almost there, but we need to pass to pinned boundary conditions, in order to take advantage of super-additivity. For this we apply Lemma 3.5 that tells us

$$Z^{\mathtt{f}}_{N,\omega,\beta,h^a_c(\beta)+h} \leq Z_{N,\omega,\beta,h^a_c(\beta)+h}\left(1 + NC(K(\cdot))\exp(-\beta\omega_N + \log \mathrm{M}(\beta) - h)\right). \tag{4.28}$$

We can of course find a value $\eta(\delta)$ such that

$$\mathbb{P}\left(\exp(-\beta\omega_N + \log \mathrm{M}(\beta) - h) \geq \eta(\delta)\right) \leq \frac{\delta}{4}, \tag{4.29}$$

and (4.27)–(4.29) imply

$$\frac{\delta}{4} \leq \mathbb{P}\left(Z_{N,\omega,\beta,h^a_c(\beta)+h} \geq \frac{\exp\left(N^{\alpha\varepsilon}\right)}{4(1+(\eta CN))}\right) \leq \mathbb{P}\left(Z_{N,\omega,\beta,h^a_c(\beta)+h} \geq \exp\left(N^{\alpha\varepsilon}/2\right)\right), \tag{4.30}$$

where the last inequality holds for h sufficiently small (i.e. N sufficiently large). From this and by using the basic bound $Z_{N,\omega,\beta,h^a_c(\beta)} \geq K(N)\exp(\beta\omega_N - \log \mathrm{M}(\beta))$, cf. (3.17), we extract [call $\widehat{A}_N(\varepsilon)$ the event in the right-most term in (4.30)]:

$$\begin{aligned} &\mathbb{E}\log Z_{N,\omega,\beta,h_c^a(\beta)+h} \\ &\quad \geq \mathbb{E}\left[\log Z_{N,\omega,\beta,h_c^a(\beta)+h};\widehat{A}_N(\varepsilon)\right]+\log K(N)+\beta\mathbb{E}[\omega_1;\,\omega_1<0]-\log \mathrm{M}(\beta) \\ &\qquad\qquad \geq \frac{\delta}{4}\frac{N^{\alpha\varepsilon}}{2}-c\log N, \end{aligned} \tag{4.31}$$

for some $c>0$. Therefore for h sufficiently small we have

$$\mathbb{E}\log Z_{N,\omega,\beta,h_c^a(\beta)+h}>0, \tag{4.32}$$

which tells us, by super-additivity, that the model is localized for every $h>0$, that is $h_c(\beta)\leq h_c^a(\beta)$, which implies $h_c(\beta)=h_c^a(\beta)$ [as repeatedly stressed: $h_c(\beta)\geq h_c^a(\beta)$ by annealing, cf. (3.16)].

But one can go beyond: for h as above (that is smaller than a constant that depends on ε) we can also make $\mathbb{E}\log Z_{N,\omega,\beta,h_c(\beta)+h}$ larger than a fixed constant (say, one) so that (super-additivity, again!)

$$\mathrm{F}(\beta,h_c(\beta)+h)\geq\frac{1}{N}\mathbb{E}\log Z_{N,\omega,\beta,h_c(\beta)+h}\geq\frac{1}{N}\geq\frac{1}{2}h^{\frac{1}{\alpha(1-2\varepsilon)}}. \tag{4.33}$$

Since ε can be chosen arbitrarily small, we are done. □

Remark 4.8. In [9] Lemma 4.6 is circumvented by using the fact that at the annealed critical point $\{Z^{\mathrm{f}}_{N,\omega}\}_{N=0,1,\ldots}$ is a martingale with respect to the natural filtration of the sequence ω. Since it is non-negative, it converges to $Z^{\mathrm{f}}_{\infty,\omega}:=\limsup_N Z^{\mathrm{f}}_{N,\omega}$. Therefore, if $\{Z^{\mathrm{f}}_{N,\omega}\}_{N=0,1,\ldots}$ is uniformly integrable we have also convergence in L^1 and (4.21) easily follows. It is also interesting to note that $Z^{\mathrm{f}}_{\infty,\omega}>0$ is a tail event and therefore it has probability zero or one: when the sequence is uniformly integrable we therefore have $Z^{\mathrm{f}}_{\infty,\omega}>0$ $\mathbb{P}(\mathrm{d}\omega)$-a.s.. Last, but not least, if the results we presented in this chapter yield a weaker disorder irrelevance than what in proven in [1,10], the method works as soon as uniform integrability (of the partition function at annealed criticality) is established, while [1,10] are based on L^2 estimates.

4.3 A Look at the Literature

The references for the physics part of this chapter are already in the text and they will be taken up again in the next chapters.

The first proof of disorder irrelevance for $\alpha<1/2$ and $\beta<\beta_0$ is due to Alexander [1], who has actually proven that for every $\varepsilon>0$ there exists $\beta_0(\varepsilon)>0$ such that for $\beta\leq\beta_0(\varepsilon)$ we have $\liminf_{\Delta\searrow 0}\mathrm{F}(\beta,h_c^a(\beta)+\Delta)/\mathrm{F}(0,\Delta)\geq 1-\varepsilon$. Such a sharp estimate on the closeness between quenched and annealed free energies has been established for Gaussian charges. A different proof, based on spin glass techniques, of this result has been given by Toninelli [10]. A further sharpening of these results, providing in particular a rigorous version of (4.4), can be found in [7]. All these results have been proven in the extended framework of regularly varying $K(\cdot)$. Such an extended framework allows dealing with $\alpha = 0$ too: this case has the interesting feature that disorder is irrelevant for all β [2], regardless of details of $K(\cdot)$ and of the law of ω_1 (this phenomenon, that is $\beta_0 = \infty$, may come up also for $\alpha > 0$, but for suitable choices of $K(\cdot)$ and ω_1).

We point out also that recently another approach to showing $h_c(\beta) = h_c^a(\beta)$ for $\alpha < 1/2$ has been set forth in [4]. This approach is based on the quenched large deviation principle proven in [3].

References

1. K.S. Alexander, The effect of disorder on polymer depinning transitions. Commun. Math. Phys. **279**, 117–146 (2008)
2. K.S. Alexander, N. Zygouras, Equality of critical points for polymer depinning transitions with loop exponent one. Ann. Appl. Probab. **20**, 356–366 (2010)
3. M. Birkner, A. Greven, F. den Hollander, Quenched large deviation principle for words in a letter sequence. Probab. Theory Relat. Fields **148**, 403–456 (2010)
4. D. Cheliotis, F. den Hollander, Variational characterization of the critical curve for pinning of random polymers. arXiv:1005.3661
5. B. Derrida, V. Hakim, J. Vannimenus, Effect of disorder on two-dimensional wetting. J. Stat. Phys. **66**, 1189–1213 (1992)
6. G. Forgacs, J.M. Luck, Th. M. Nieuwenhuizen, H. Orland, Wetting of a disordered substrate: exact critical behavior in two dimensions. Phys. Rev. Lett. **57**, 2184–2187 (1986)
7. G. Giacomin, F.L. Toninelli, On the irrelevant disorder regime of pinning models. Ann. Probab. **37**, 1841–1873 (2009)
8. A.B. Harris, Effect of Random defects on the critical behaviour of Ising models. J. Phys. C **7**, 1671–1692 (1974)
9. H. Lacoin, The martingale approach to disorder irrelevance for pinning models. Electron. Commun. Probab. **15**, 418–427 (2010)
10. F.L. Toninelli, A replica-coupling approach to disordered pinning models. Commun. Math. Phys. **280**, 389–401 (2008)

Chapter 5
Relevant Disorder Estimates: The Smoothing Phenomenon

Abstract We show that, for $\alpha > 1/2$ and as soon as $\beta > 0$, disorder is relevant, in the sense that the critical behavior of the disordered system differs from the one of the pure, i.e. homogeneous, system. We do this by establishing a *smoothing inequality* for the free energy. We then review the literature on the effect of the disorder on phase transitions. In doing so we will present a number of physical predictions on disordered Ising models that are challenges for mathematicians.

5.1 Smoothing for Gaussian Charges: The Rare Stretch Strategy

We are going to prove the following.

Theorem 5.1. *If* $\omega_1 \sim \mathcal{N}(0,1)$ *then*

$$\textsc{f}(\beta, h_c(\beta) + \Delta) \leq c \frac{(1+\alpha)}{\beta^2} \Delta^2, \tag{5.1}$$

with $c = (e - (1/2))/(e-1)$, *for every* β *and* Δ.

Of course this result is trivial if $\beta = 0$ or if $\Delta \leq 0$. But it does say something relevant if $\beta > 0$ and if $\Delta > 0$ and to make this more clear let us rewrite (5.1) as

$$0 \leq \textsc{f}(\beta, h_c(\beta) + \Delta) - \textsc{f}(\beta, h_c(\beta)) \leq c \frac{(1+\alpha)}{\beta^2} \Delta^2, \tag{5.2}$$

so that it becomes evident that it is an estimate on the modulus of continuity of the free energy in the h variable. This bound is therefore telling us that, at criticality, the free energy of the quenched system is smoother than the free energy of the pure

G. Giacomin, *Disorder and Critical Phenomena Through Basic Probability Models*, Lecture Notes in Mathematics 2025, DOI 10.1007/978-3-642-21156-0_5,

© Springer-Verlag Berlin Heidelberg 2011

system for $\alpha > 1/2$ (recall Theorem 2.10). In terms of critical exponents, what we retain of (5.2) is

$$\liminf_{\Delta \searrow 0} \frac{\log \mathrm{F}(\beta, h_c(\beta)+\Delta)}{\log \Delta} \geq 2. \tag{5.3}$$

Proof. Let us fix a value of $\beta > 0$. In what follows $h \geq h_c(\beta)$, but in fact in the end we will set $h = h_c(\beta)$. Moreover ℓ is a large positive integer and we assume $N/\ell \in \mathbb{N}$. We introduce the sequence of IID Bernoulli random variables $X = \{X_j\}_{j=1,2,\ldots}$ defined as the indicator functions of the sets

$$\left\{\omega : \log Z_{\ell-1,\theta^{j\ell+1}\omega} \geq (1-\varepsilon)\ell \mathrm{F}(\beta, h+\Delta) \text{ and } \omega_{j\ell+1} \geq 0\right\}, \tag{5.4}$$

where $\Delta > 0$, so that $\mathrm{F}(\beta, h+\Delta) > 0$, and $\varepsilon \in (0,1)$ (in the end we will send ε to zero). Therefore the parameter of the Bernoulli variables is

$$p(\ell) := \mathbb{P}\left(\log Z_{\ell-1,\omega} \geq (1-\varepsilon)\ell \mathrm{F}(\beta, h+\Delta),\, \omega_0 \geq 0\right). \tag{5.5}$$

By the self-averaging property of the free energy (Proposition 3.2) we have that $p(\ell) = o_\ell(1)$ and it isn't too difficult to see that it is even exponentially small. But we claim that

$$\liminf_{\ell \to \infty} \frac{1}{\ell} \log p(\ell) \geq -\frac{\Delta^2}{2\beta^2}. \tag{5.6}$$

A proof of this bound can be found below, but we want to point out that such a result should not come as a surprise: we give now a heuristic argument for it in which at times we use ℓ instead of $\ell - 1$, but this is of course inessential in the limit: the choice of reducing the partition function to size $\ell - 1$ is to have full independence between what happens in the blocks, but there are plenty of ways to get around this minor point. The heuristics goes as follows: $\sum_{j=1}^{\ell} \omega_j \approx \ell\Delta/\beta$ is a large deviation event of probability about $\exp(-\ell\Delta^2/(2\beta^2))$ for ℓ large. On such an event the first ℓ variables of the sequence look like $\omega_1 + \Delta/\beta, \ldots, \omega_\ell + \Delta/\beta$ so that, when such an event occurs, the system in the block $\{1, \ldots, \ell\}$ looks like the original system with h replaced by $h + \Delta$: (5.6) is therefore plausible, since almost surely $\ell^{-1} \log Z_{\ell,\omega,\beta,h+\Delta} \longrightarrow \mathrm{F}(\beta, h+\Delta)$.

We now introduce the *good set* of charges by setting

$$\iota_N(\omega) := \inf\left\{j \in \mathbb{N} \cup \{0\} : X_j(\omega) = 1\right\} \text{ and } G_{N,\ell} := \left\{\omega : \iota_N(\omega) < \frac{N}{\ell} - 1\right\}, \tag{5.7}$$

where we have insisted on $\iota_N(\omega) < (N/\ell) - 1$ instead of $\iota_N(\omega) < N/\ell$ to have $G_{N,\ell}$ independent of ω_N. For $\omega \in G_{N,\ell}$, we make a lower bound on the partition function by selecting the renewal trajectories that visit only the $\iota_N(\omega)$th block, more precisely

$$Z_{N,\omega} \geq Z_{N,\omega}\left(\tau_1 = \iota_N(\omega)\ell + 1,\, \tau \cap [(\iota_N(\omega)+1)\ell, N) = (\iota_N(\omega)+1)\ell\right). \tag{5.8}$$

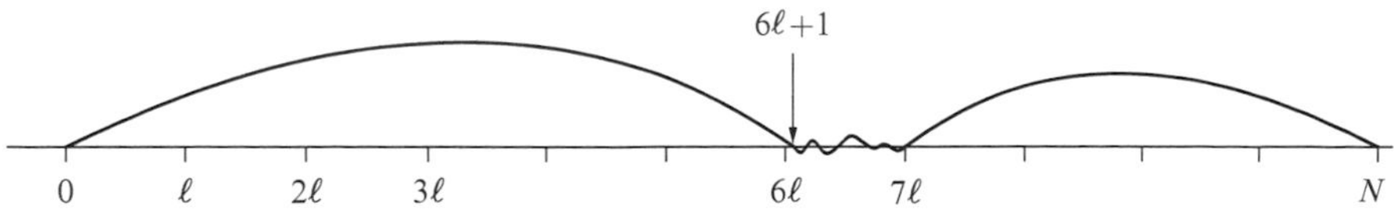

Fig. 5.1 One of the trajectories selected to obtain a lower bound on the partition function: in this case $N = 11\ell$ and $\iota_N(\omega) = 6$. In the proof N is chosen as a function of ℓ in such a way to guarantee that at least one of the events (5.4) is realized with probability bounded away from 0 for all ℓ

In Fig. 5.1 we have drawn a case in which $N = 11\ell$ and $\iota_N(\omega) = 6$. The point is that the right-hand side of (5.8) can be expressed explicitly by using the renewal property, but there is still an important point to take care of: we want $\mathbb{P}(G_{N,\ell})$ bounded away from zero and this requires N sufficiently large, because $\mathbb{P}(G_{N,\ell}) = 1-(1-p(\ell))^{-1+(N/\ell)}$. So we choose

$$N = N(\ell) := \ell \left\lceil \frac{1}{p(\ell)} + 1 \right\rceil, \tag{5.9}$$

so that N is bounded below by $\ell/p(\ell)$ (but it is also arbitrarily close to it when ℓ is sufficiently large!) and $\mathbb{P}(G_{N,\ell}) \geq 1-e^{-1}$.

The computation is now rather straightforward: the right-hand side of (5.8) is equal to

$$K(\iota_N(\omega)\ell+1)\exp(\beta\omega_{\iota_N(\omega)\ell+1}+h) \\ \times Z_{\ell-1,\theta^{\iota_N(\omega)\ell+1}\omega} K(N-(\iota_N(\omega)+1)\ell)\exp(\beta\omega_N+h), \tag{5.10}$$

and by using $X_{\iota_N(\omega)} = 1$ and (2.30) we obtain that ($C_1(K(\cdot)) > 0$)

$$Z_{N,\omega} \geq \frac{C_1(K(\cdot))}{N^{2(1+\alpha)}} e^h \exp\left((1-\varepsilon)\ell \mathrm{F}(\beta,h+\Delta)\right)\exp(\beta\omega_N+h). \tag{5.11}$$

Therefore [use the independence of ω_N and $G_{N,\ell}$ and recall (5.9)]

$$\begin{aligned} &\mathbb{E}\left[\log Z_{N,\omega};\, G_{N,\ell}\right] \\ &\quad\geq (\log(C_1)+2h-2(1+\alpha)\log N+(1-\varepsilon)\ell\mathrm{F}(\beta,h+\Delta))\,\mathbb{P}(G_{N,\ell}) \\ &\quad\geq \left(-2\frac{1+\alpha}{1-\varepsilon}\log\left(\frac{\ell}{p(\ell)}\right)+(1-\varepsilon)\ell\mathrm{F}(\beta,h+\Delta)\right)\mathbb{P}(G_{N,\ell}), \end{aligned} \tag{5.12}$$

where the factor $(1-\varepsilon)$ in the denominator of the first term in the right-hand side takes care, for ℓ sufficiently large, of replacing the correct value of N by $\ell/p(\ell)$ and of the $O(1)$ terms. On the other hand by using the bound in Remark 3.3 we directly get ($C_2(K(\cdot) > 0$)

$$\begin{aligned}\mathbb{E}\left[\log Z_{N,\omega}; G_{N,\ell}^{\complement}\right] &\geq \left(\log \frac{C_2(K(\cdot))}{N^{1+\alpha}} + h\right)\left(1 - \mathbb{P}(G_{N,\ell})\right) \\ &\geq \left(-\frac{1+\alpha}{1-\varepsilon}\log\left(\frac{\ell}{p(\ell)}\right)\right)\left(1 - \mathbb{P}(G_{N,\ell})\right).\end{aligned} \tag{5.13}$$

Let us now put (5.12) and (5.13) together keeping into account that (5.6) guarantees that

$$\log\left(\frac{\ell}{p(\ell)}\right) \leq \ell \frac{\Delta^2}{2\beta^2(1-\varepsilon)}, \tag{5.14}$$

for ℓ sufficiently large. We therefore get to

$$\mathbb{E}\log Z_{N,\omega} \geq -(1+\mathbb{P}(G_{N,\ell}))\frac{(1+\alpha)}{(1-\varepsilon)^2}\ell\frac{\Delta^2}{2\beta^2} + \mathbb{P}(G_{N,\ell})(1-\varepsilon)\ell \mathrm{F}(\beta, h+\Delta). \tag{5.15}$$

It is now the moment to recall that $\{\mathbb{E}\log Z_{N,\omega}\}_N$ is super-additive, cf. (3.6), and to use (or set) $h = h_c(\beta)$ so that for every N

$$0 = \mathrm{F}(\beta, h_c(\beta)) \geq \frac{1}{N}\mathbb{E}\log Z_{N,\omega}, \tag{5.16}$$

that entails the non-positivity of the right-hand side of (5.15) and therefore

$$\mathrm{F}(\beta, h_c(\beta)+\Delta) \leq \frac{1+\mathbb{P}(G_{N,\ell})}{2\mathbb{P}(G_{N,\ell})}\frac{(1+\alpha)\Delta^2}{(1-\varepsilon)^3\beta^2}, \tag{5.17}$$

and by using $\mathbb{P}(G_{N,\ell}) \geq 1 - e^{-1}$ and by taking ε arbitrarily small we are done. □

Proof of (5.6). Without loss of generality we set $j = 0$ in (5.4) and we point out that, since ω_0 is independent of $Z_{\ell-1,\omega}$, requiring $\omega_0 \geq 0$ (an event of probability $1/2$) has no effect on the estimate we are after, so we neglect it. Moreover we prove the statement for $Z_{\ell,\omega}$ instead of $Z_{\ell-1,\omega}$ for ease of notation. For $a \in \mathbb{R}$ call $\mathbb{P}_{\ell,a}$ the law of $(\omega_1 + a, \omega_2 + a, \ldots, \omega_\ell + a)$. Call E_ℓ the event in $\sigma(\omega_1, \ldots, \omega_\ell)$ of which we want to estimate the probability and note that $\lim_\ell \mathbb{P}_{\ell,\Delta/\beta}(E_\ell) = 1$ by Proposition 3.2, that is E_ℓ becomes *typical* under the new measure for ℓ large. It is not difficult to compute the relative entropy of $\mathbb{P}_{\ell,0}$ with respect to $\mathbb{P}_{\ell,\Delta/\beta}$:

$$\mathscr{H}\left(\mathbb{P}_{\ell,\Delta/\beta} \middle| \mathbb{P}_{\ell,0}\right) := \mathbb{E}_{\ell,\Delta/\beta}\left[\log\frac{\mathrm{d}\mathbb{P}_{\ell,\Delta/\beta}}{\mathrm{d}\mathbb{P}_{\ell,0}}\right] = \ell\frac{\Delta^2}{2\beta^2}. \tag{5.18}$$

But (we use $\mathbb{P} = \mathbb{P}_{\ell,0}$ and $\widetilde{\mathbb{P}} = \mathbb{P}_{\ell,\Delta/\beta}$ for readability)

$$
\begin{aligned}
\log \frac{\mathbb{P}(E_\ell)}{\widetilde{\mathbb{P}}(E_\ell)} &= \log \widetilde{\mathbb{E}}\left[\left.\frac{\mathrm{d}\mathbb{P}}{\mathrm{d}\widetilde{\mathbb{P}}}\right| E_\ell\right] \geq \widetilde{\mathbb{E}}\left[\log \left.\frac{\mathrm{d}\mathbb{P}}{\mathrm{d}\widetilde{\mathbb{P}}}\right| E_\ell\right] \\
&= -\frac{1}{\widetilde{\mathbb{P}}(E_\ell)}\mathbb{E}\left[\frac{\mathrm{d}\widetilde{\mathbb{P}}}{\mathrm{d}\mathbb{P}}\log\frac{\mathrm{d}\widetilde{\mathbb{P}}}{\mathrm{d}\mathbb{P}}; E_\ell\right] \geq -\frac{1}{\widetilde{\mathbb{P}}(E_\ell)}\mathbb{E}\left[\frac{\mathrm{d}\widetilde{\mathbb{P}}}{\mathrm{d}\mathbb{P}}\log\frac{\mathrm{d}\widetilde{\mathbb{P}}}{\mathrm{d}\mathbb{P}}+\frac{1}{e}\right] \\
&= -\frac{1}{\widetilde{\mathbb{P}}(E_\ell)}\left(\mathscr{H}\left(\widetilde{\mathbb{P}}\middle|\mathbb{P}\right)+\frac{1}{e}\right), \qquad (5.19)
\end{aligned}
$$

where we have used Jensen inequality and the fact that $x\log x + 1/e \geq 0$ for every $x > 0$. Plug (5.18) into the last estimate and use the observation we made just before (5.18) to complete the proof. □

5.2 More General Charge Distributions

One can upgrade Theorem 5.1 to more general charge distributions.

Theorem 5.2. *If in addition to Hypothesis 3.1 we assume that ω_1 is such that there exists $c > 0$ such that relative entropy of the law of $\omega_1 + x$ with respect to the law of ω_1 is bounded by cx^2 for every $x \in \mathbb{R}$, then for every $\beta > 0$ there exists $C(\beta) > 0$ (depending of course also on the law of ω_1) such that*

$$
\mathrm{F}(\beta, h_c(\beta)+\Delta) \leq C(\beta)(1+\alpha)\Delta^2, \qquad (5.20)
$$

for every $\beta > 0$ and every Δ. One can choose $C(\beta)$ such that $\beta^2 C(\beta)$ stays bounded as $\beta \searrow 0$.

It is easily verified that the second condition holds for example when ω_1 has a density and such a density can be written as $\exp(-V(\cdot))$ with $V(\cdot)$ a polynomial which is bounded below.

Proof. The proof is just an exercise in modifying (5.18). □

Remark 5.3. Theorem 5.2 generalizes to the case of bounded charges, that is $\mathbb{P}(|\omega_1| > c) = 0$ for some $c > 0$. In this case the proof goes through via tilting, rather than shifting, the charge distribution, see [23].

5.3 Back to and Beyond Harris Criterion: Disorder and Smoothing

The Harris criterion or, more generally, Harris type arguments are not of much help in saying what happens if disorder is relevant. But this is the occasion to reconsider Harris' strategy from the start and the natural framework is the original [26] one: diluted Ising model.

5.3.1 Disorder and Phase Transitions

Early approaches to quenched disorder were dominated by the idea that disorder destroys critical points (see for example [18, 29] and references therein) and the general rationale behind this was that a disordered system should break up into differently behaved systems, in particular every one with its own critical point which overall results in a general smoothing of singularities. But later on it became clear that some transitions survive to switching disorder on and Harris' approach [26] took a leading stand that was and still is analyzed, generalized and criticized by several authors, e.g. [20, 29]. Recently moreover, these ideas have been taken up and applied also to quantum mechanical models and to non-equilibrium transitions with the discovery of very rich behaviors that are only partially understood, but we will not develop this point beyond referring to [22, 25, 29] as starting point for references.

Central to the Harris approach is the notion of *weak disorder*. It is somewhat customary not to talk about weak disorder when dealing with frustration or with external fields. What is left contains systems with spatial variations of the coupling potentials and this is notably the case of the Ising model, see Sect. 2.5 of Chap. 2, with $h(\cdot) = \text{const.}$ (but in fact we have above all in mind $h(\cdot) = 0$, that is no external field) and random $J(x,y)$ (say IID: the sequence is indexed by the unordered couple $\{x,y\}$): the diluted Ising model just corresponds to $J(x,y) \in \{0,1\}$, that is the coupling potentials are just $B(1-p)$ variables with $p \in [0,1]$: $p = 0$, *no defects*, is the homogeneous (or *pure*) model and $p = 1$ is just the trivial (uncorrelated) spin model. Harris' disorder irrelevance criterion in this framework boils down to $\nu > 2/d$ (or to $\nu \geq 2/d$, we'll come back to this below), where ν is the critical exponent of the correlation length for the pure system and d is the dimension. Two observations are now in order:

1. In the original work Harris does consider mostly the diluted Ising model, but he treats also, in a more expedite way, what he calls *magnetic glass*, which is an Ising model with $J(\cdot,\cdot)$ IID, of mean $J > 0$ and small variance (the small parameter is precisely the variance).
2. It is customary to say that Harris criterion says that the disorder is irrelevant if the critical exponent $\widetilde{\alpha}$ of the specific heat, that is the second derivative of the free energy with respect to β for the Ising model (that should then behave like $|\beta - \beta_c|^{-\widetilde{\alpha}}$), is negative: it is not customary to put a tilde over the exponent, but α in this work is reserved for the inter-arrival law. The *hyperscaling relation* [21]

$$d\nu = 2 - \widetilde{\alpha}, \tag{5.21}$$

reconciles this alternative viewpoint on the Harris criterion with the previous one ($\nu > 2/d$), but it should be noticed that, for the time being, in a reasonable generality it has been shown only that $d\nu \geq 2 - \widetilde{\alpha}$ and the equality is not expected to hold (and in some cases we can say that it does not hold) for $d \geq d_c$, with d_c the *upper critical dimension*, i.e. the dimension starting from which the critical exponents become the mean field ones ($d_c = 4$ for the Ising model [21]).

5.3.2 Harris' Heuristic Argument

Harris' original work [26] is at times reduced to a heuristic argument on critical exponents that we are going to reproduce below, but in reality it is mostly dedicated to a perturbative expansions of the free energy and to *arguing* that certain terms are the leading contribution to each order: both the dependence of the critical temperature on the disorder intensity (p) and the identification of the critical exponent are addressed. Harris then justifies his approach in more intuitive/heuristic way and this goes (more or less) as follows. First of all we apply the rather convincing idea (though not always easy to be translated into a rigorous argument) that an infinite system may be seen as a collection of approximately independent systems of correlation length size $\kappa_p = \kappa_p(\beta)$ (β is the inverse temperature). We assume that $\kappa_p \approx |\beta - \beta_c(p)|^{-\nu(p)}$, so the exponent ν used above is $\nu(0)$: Harris assumes that $\nu(p)$ behaves smoothly at least for small p. Now, the typical number of defects in a volume of correlation length size is $\kappa_p^d p$ and the variance of this quantity is $\kappa_p^d p(1-p)$, so that the standard deviation of the density of defects is $\kappa_p^{-d/2}\sqrt{p(1-p)}$. So the question is: what is the critical temperature for a system with such a density of defects? Harris argues that the difference $\beta_c(p) - \beta_c(0)$ between disordered and pure critical points is, to leading order, linear in p (this, regularity apart, amounts to saying that $\beta_c'(0) \neq 0$), so that the critical point in the box we are considering deviates from the *typical* one $\beta_c(p)$ of a quantity that is proportional to $\kappa_p^{-d/2}\sqrt{p(1-p)}$. If now we write the scaling behavior of the correlation length near criticality as

$$|\beta - \beta_c| \approx \kappa_p(\beta)^{-1/\nu(p)}, \tag{5.22}$$

we see that the disorder fluctuations are negligible only if

$$\kappa^{-d/2}(\beta)\sqrt{p(1-p)} \ll \kappa_p(\beta)^{-1/\nu(p)}, \tag{5.23}$$

that is if $\nu(p) > 2/d$. If $\nu(p) = \nu(0)$ then to the inequality we have *obtained* becomes the Harris criterion for disorder irrelevance, which is therefore a *necessary* condition for disorder irrelevance, that is for $\nu(p) = \nu(0)$. We could also go a bit beyond and say that, being $p(1-p)$ small, there is room for (5.22) to hold also for $\nu(p) \geq 2/d$, and $\nu \geq 2/d$ is at times considered to be the Harris criterion, however Harris himself was aware from the start of the delicate character of such an enlarged statement). Note that however the argument we gave just above suggests that $\nu(p) > 2/d$ or at least that $\nu(p) \geq 2/d$ should hold in *general* and this is in fact the bound that one can find in [13, 14].

5.3.3 Relevance and Irrelevance

Now two scenarios are in front of us:

- Either disorder is irrelevant, that is, coarse graining the system suppresses the disorder and the system on large scale is essentially homogeneous (and the critical exponents coincide with the ones of the pure system): this is expected if $\nu > 2/d$ and for moderate values of p
- Or disorder is relevant, that is, it grows under coarse graining and the system remains inhomogeneous on all length scales. Still, in this case at least two different scenarios have been put in evidence, in the sense that
  - Either the system inhomogeneities grow in a *nice* way and the system can be *(re)normalized* to obtain a limit model that is disordered, but the disorder is finite and one should observe conventional power law scaling approaching criticality. This is possibly the behavior expected for pinning models in the relevant disorder regime [16]. It is interesting to observe that in the physical literature it is conjectured that at and near criticality suitably rescaled macroscopic observables are not self-averaging [1, 31] (to the author's knowledge there is no model for which such a behavior has been established rigorously and even no example in which the new critical exponent has been identified).
  - Or the coarse grained system is still so strongly inhomogeneous that the fixed points of such a renormalization procedure have *infinite* disorder. This type of behavior has been established, thanks to the celebrated solution of McCoy and Wu [27], in the two dimensional Ising model with disorder which is completely correlated in the direction of one of the axes. But infinite disorder renormalization fixed points, i.e limit models, appear to be rather ubiquitous in quantum transitions (e.g. [22, 28]).

5.3.4 The Diluted Ising Model

But, in the end, what really happens to the diluted Ising model without external field when disorder is introduced? Of course rigorous results can be expected only in dimension two, for which an exact solution is available due to Onsager in 1944 (see e.g. [4]), and in sufficiently large dimension for which we know rigorously that the critical exponents coincide with the mean field ones [21], but the critical exponents of the pure Ising model are considered to be known with a good precision in the physical literature, so that we can discuss the issue in all dimensions. And it turns out that:

- In $d = 2$ the second derivative of the free energy of the pure Ising model diverges like $\log(1/|\beta - \beta_c|)$ at criticality, that is $\widetilde{\alpha} = 0$: but this means that $\nu = 2$ (hyperscaling) and we are precisely in the so called *marginal* case in which the Harris criterion for irrelevance is inconclusive (as a matter of fact Harris dwells at length on this in his original paper [26]). In a subsequent work [17] it is claimed that in the diluted two dimensional Ising model the second derivative of the free energy near criticality behaves like $\log\log(1/|\beta-\beta_c|)$ and therefore the critical exponent is not changed: it is therefore natural to call such a case *marginally irrelevant*. It should be however remarked that in spite of the result we just stated, the same authors claim [18] also that the correlations at $\beta=\beta_c$ and for p small decay like $\exp(-cp^{-1}\log\log r)$, where c is a positive constant and r is the distance of the two spins of which we are considering the correlation. This means that the exponent is zero, while the same quantity for the pure system decays with exponent $1/4$, that is like $1/r^{1/4}$.

- In $d=3$ one expects the critical exponent ν of the free energy of the pure system to be $0.627\ldots<2/d$, so one expects disorder to be relevant and it is claimed that the new critical exponent is $0.684\ldots>2/d$ (plenty of literature on this point, with what seems to be a fairly good agreement, at least on the general picture: the numerical data are taken from [29]).
- It is substantially harder to find literature for d above or at the upper critical dimension $d_c=4$. But the exponent ν for the mean field case is $1/2$, so that $\nu>2/d$ for $d>4$ (irrelevance!) and one has equality at the upper critical dimension $d=4$ (a delicate case also for the pure model! [21]). What we have just stated is in agreement with [30] where the Harris criterion is generalized to correlated disorder. Note on the way that $\widetilde{\alpha}$ in the mean field case is zero [21].

5.3.5 Random External Fields

At least a word is absolutely due to the case of random external fields and, for conciseness, let us restrict to the Ising model with $J(x,y)=J>0$ for every x and y and $\{h(x)\}_x$ IID (say, standard Gaussian). For a model in a box of linear size N we can argue, in a very rough way but essentially following Imry and Ma, that the boundary spins (say, all up) directly affect (a constant times) N^{d-1} spins, while the external field acts on N^d spins, but in the case of centered variables it accounts only for a net effect of the order of $\sqrt{N^{d/2}}$. So the two effects are really competing in dimension two, while starting from dimension three one tends to say that the boundary effect takes over and a phase transition is still observed (this is in fact the case, but the issue was debated for a long time in physics, before and even after a rigorous proof was established [10]). But let us focus on the $d=2$ case because in this case pushing farther the Imry–Ma argument one tends to argue that even a very small amount of IID disorder does destroy the phase transition: this is essentially due to the fact that while boundary conditions are fixed, in the bulk the disorder can become occasionally very large, in the sense that the net effect that we were mentioning above of $\sqrt{N^{d/2}}=N$ should be rather seen as N times a Gaussian random variable, that can take arbitrarily large values. This has been made rigorous by Aizenman and Wehr [2] who showed that in this case the disorder completely smears off the transition: the disordered model (in two dimensions) does not have a phase transition. We remark here that in the pinning case the boundary plays no role and there appears not to be an analog of the Imry–Ma/Aizenman–Wehr arguments. And in fact the argument that we have developed in this chapter is substantially different from the one in [2].

5.4 A Further Look at the Literature

The smoothing technique detailed in this chapter has been introduced in [23,24] and it has been taken up in other contexts, notably for pinning of directed semi-flexible polymers [11] and, in [5], for the *pinning on a walk* model introduced in [6]. One of the ingredients of the proof is identifying rare regions in the environment in which the process can obtain an atypically large energetic contribution: with respect to this, the original proof was going through the argument set forth in [7] for the *copolymer in selective solvents* model, while here we present an argument that exploits the super-additive property of the model, like in [5], at the very cheap expense of a poorer multiplicative constant c in (5.1) ($c = 1/2$ in the original approach). It should be noted however that such a loss in the constant is not minor for copolymer models [7, 8] and a more involved argument, but based on super-additivity and close in spirit to what we present here, without loss in the constant can be found in [12]. Smoothing has been shown to be absent in some models with $K(n)$ decaying exponentially fast [3, 15].

For the second part of this chapter, dealing with Harris' approach, we would only like to stress again the lack of convincing heuristics, not to speak of rigorous results, on the critical behavior of disordered pinning models (and not only!) when disorder is relevant.

References

1. A. Aharony, A.B. Harris, Absence of self-averaging and universal fluctuations in random systems near critical points. Phys. Rev. Lett. **77**, 3700–3703 (1996)
2. M. Aizenman, J. Wehr, Rounding effects of quenched randomness on first-order phase transitions. Commun. Math. Phys. **130**, 489–528 (1990)
3. K.S. Alexander, Ivy on the ceiling: first-order polymer depinning transitions with quenched disorder. Markov Process. Relat. Fields **13**, 663–680 (2007)
4. R.J. Baxter, *Exactly Solved Models in Statistical Mechanics* (Academic, London, 1982)
5. Q. Berger, H. Lacoin, The effect of disorder on the free-energy for the random walk pinning model: smoothing of the phase transition and low temperature asymptotics. J. Statist. Phys. **42**, 322–341 (2011)
6. M. Birkner, R. Sun, Annealed vs quenched critical points for a random walk pinning model. Ann. Inst. H. Poincaré (B) Probab. Stat. **46**, 414–441 (2010)

7. T. Bodineau, G. Giacomin, On the localization transition of random copolymers near selective interfaces. J. Stat. Phys. **117**, 801–818 (2004)
8. T. Bodineau, G. Giacomin, H. Lacoin, F.L. Toninelli, Copolymers at selective interfaces: new bounds on the phase diagram. J. Stat. Phys. **132**, 603–626 (2008)
9. A. Bovier, C. Külske, There are no nice interfaces in $(2+1)$-dimensional SOS models in random media. J. Stat. Phys. **83**, 751–759 (1996)
10. J. Bricmont, A. Kupiainen, Phase transition in the 3d random field Ising model. Commun. Math. Phys. **116**, 539–572 (1988)
11. F. Caravenna, J.-D. Deuschel, Scaling limits of (1+1)-dimensional pinning models with Laplacian interaction. Ann. Probab. **37**, 903–945 (2009)
12. F. Caravenna, G. Giacomin, M. Gubinelli, A numerical approach to copolymers at selective interfaces. J. Stat. Phys. **122**, 799–832 (2006)
13. J.T. Chayes, L. Chayes, D.S. Fisher, T. Spencer, Finite-size scaling and correlation lengths for disordered systems. Phys. Rev. Lett. **57**, 2999–3002 (1986)
14. J.T. Chayes, L. Chayes, D.S. Fisher, T. Spencer, Correlation length bounds for disordered Ising ferromagnets. Commun. Math. Phys. **120**, 501–523 (1989)
15. M. Cranston, O. Hryniv, S. Molchanov, Homo- and hetero-polymers in the mean-field approximation. Markov Process. Relat. Fields **15**, 205–224 (2009)
16. B. Derrida, E. Gardner, Renormalization group study of a disordered model. J. Phys. A Math. Gen. **17**, 3223–3236 (1984)
17. Vik. S. Dotsenko, Vl. S. Dotsenko, Critical behavior of the 2D Ising model with impurity bonds. J. Phys. C Solid State Phys. **15**, 495–507 (1982)
18. Vik. S. Dotsenko, Vl. S. Dotsenko, Critical behaviour of the phase transition in the 2D Ising Model with impurities. Adv. Phys. **32**, 129–172 (1983)
19. Vik. S. Dotsenko, A.B. Harris, D. Sherrington, R.B. Stinchcombe, Replica-symmetry breaking in the critical behaviour of the random ferromagnet. J. Phys. A **28**, 3093–3107 (1995)
20. Vik. S. Dotsenko, On the nature of the phase transition in the three-dimensional random field Ising model. J. Stat. Mech., P09005 (2007)
21. R. Fernández, J. Frölich, A.D. Sokal, *Random Walks, Critical Phenomena, and Triviality in Quantum Field Theory*. Texts and Monographs in Physics (Springer, New York, 1992)
22. D.S. Fisher, Random transverse field Ising spin chains. Phys. Rev. Lett. **69**, 534–537 (1992)
23. G. Giacomin, F.L. Toninelli, Smoothing effect of quenched disorder on polymer depinning transitions. Commun. Math. Phys. **266**, 1–16 (2006)
24. G. Giacomin, F.L. Toninelli, *Smoothing of depinning transitions for directed polymers with quenched disorder*. Phys. Rev. Lett **96**, 070602 (2006)
25. R.L. Greenblatt, M. Aizenman, J.L. Lebowitz, On spin systems with quenched randomness: classical and quantum. Physica A **389**, 2902–2906 (2010)
26. A.B. Harris, Effect of random defects on the critical behaviour of Ising models. J. Phys. C **7**, 1671–1692 (1974)
27. B. McCoy, T.T. Wu, *The two-dimensional Ising model* (Harvard University Press, Cambridge, MA, 1973)
28. O. Motrunich, S.-C. Mau, D.A. Huse, D.S. Fisher, Infinite-randomness quantum Ising critical fixed points. Phys. Rev. B **61**, 1160–1172 (2000)
29. T. Vojta, R. Sknepnek, Critical points and quenched disorder: from Harris criterion to rare regions and smearing. Phys. Stat. Sol. B **241**, 2118–2127 (2004)
30. A. Weinrib, B.I. Halperin, Critical phenomena in systems with long-rage-correlated quenched disorder. Phys. Rev. B **27**, 413–427 (1983)
31. S. Wiseman, E. Domany, Finite-size scaling and lack of self-averaging in critical disordered systems. Phys. Rev. Lett. **81**, 22–25 (1998)

Chapter 6
Critical Point Shift: The Fractional Moment Method

Abstract This chapter is devoted to showing that, when $\alpha \geq 1/2$, quenched and annealed critical points are different for every $\beta > 0$, with explicit estimates on the difference. Such a result follows from upper bounds on the free energy that are obtained by estimating fractional moments (of order less than one) of the partition function. Estimates for every $\beta > 0$, notably for arbitrarily small values of β, are obtained by using a change of measure argument on the law of the disorder and by coarse graining techniques. Proving such estimates becomes harder and harder as α approaches $1/2$, i.e. the *marginal disorder* case in the Harris' sense: for $\alpha = 1/2$ the Harris criterion yields no prediction and whether quenched and annealed critical points differed or not has been a debated issue in the physical literature.

6.1 Main Result and Overview

As mentioned in the last section of the previous chapter, Harris himself in [23] has been first interested in the shift of the critical point when disorder is introduced. We have seen in Chap. 4 that for pinning models with $\alpha < 1/2$ and disorder not too large quenched and annealed critical points coincide: this means that introducing the disorder does change the critical point, but precisely to match the annealed critical point. In this chapter we are going to show that if $\alpha \in [1/2, 1]$ the quenched and annealed critical points do coincide to leading order as $\beta \to 0$, but they do not coincide and, if $\alpha > 1$, the quenched and annealed critical points are not even asymptotically equivalent. We will show also that, under some conditions on the law of the disorder, that for β sufficiently large quenched and annealed critical points differ, regardless of the value of α.

Let us recall that $h_c^a(\beta)$ is the critical point of the annealed system, that is $h_c^a(\beta) = -\log \mathbf{P}(\tau_1 < \infty) + \log \mathrm{M}(\beta)$. Without loss of generality we will assume in most of the proofs that τ is recurrent, but let us state the main result of this chapter in full generality.

G. Giacomin, *Disorder and Critical Phenomena Through Basic Probability Models*, Lecture Notes in Mathematics 2025, DOI 10.1007/978-3-642-21156-0_6,
© Springer-Verlag Berlin Heidelberg 2011

Theorem 6.1. *Consider a model based on a $K(\cdot)$-renewal with $K(\cdot)$ as in* (2.30) *and with disorder distribution $\mathbb{P}$ that satisfies Hypothesis 3.1. For any $\beta_* > 0$ there exists $\mathtt{c} = \mathtt{c}(K(\cdot), \beta_*, \mathbb{P})$ such that for every $\beta \in (0, \beta_*]$ we have*

$$h_c(\beta) - h_c^a(\beta) \geq \begin{cases} \mathtt{c}\beta^2 & \text{if } \alpha > 1, \\ \mathtt{c}\beta^2/(\log(1+1/\beta))^2 & \text{if } \alpha = 1, \\ \mathtt{c}\beta^{2\alpha/(2\alpha-1)} & \text{if } \alpha \in (1/2, 1), \\ \exp\left(-1/(\mathtt{c}\beta^4)\right) & \text{if } \alpha = 1/2. \end{cases} \tag{6.1}$$

The proof of such a result is of growing complexity as α decreases. In a sense, such a complexity is piecewise constant, with discontinuities marked by the four regimes in (6.1). All the same, the arguments have a common denominator that can be explained in the following way:

1. *Fractional moment estimates on the partition function.* One can easily see that if $\sup_N \mathbb{E}[Z_{N,\omega}^\gamma] < \infty$ for a $\gamma > 0$ then $\mathtt{F}(\beta,h) = 0$, and therefore $h_c(\beta) \geq h$. This is what we have already used repeatedly if $\gamma = 1$, because $\mathbb{E}[Z_{N,\omega}]$ is fully under control: there is of course no reason to try to bound $\mathbb{E}[Z_{N,\omega}^\gamma]$ for $\gamma > 1$ because, for the purpose of establishing $\mathtt{F}(\beta,h) = 0$, is less efficient than the annealed bound ($\gamma = 1$). We are left with $\gamma \in (0,1)$: $\mathbb{E}[Z_{N,\omega}^\gamma]$ may be uniformly bounded also if the annealed bound is not (and we will find cases in which this is true). The trouble is that a direct evaluation of $\mathbb{E}[Z_{N,\omega}^\gamma]$ with $\gamma \in (0,1)$ is not evident: we will actually bound $\mathbb{E}[Z_{N,\omega}^\gamma]$ from above rather than compute it. And such bounds will be obtained in two steps [points (2) and (3) below].
2. *Estimates up to systems of correlation length sizes: change of measure estimates.* We will present a method to bound $\mathbb{E}[Z_{N,\omega}^\gamma]$ for $N \leq k$, where k is the correlation length of the annealed system. It is based on a change of measure argument and involves guessing a new law of the charges for which the annealed partition function is bounded: the annealed partition function ($\gamma = 1$: but with respect to the new law of the disorder!) is re-obtained from the fractional moment via a Hölder inequality argument that makes appear also the price we have to pay for such a change of measure. This price is too large beyond the correlation length size and the change of measure argument becomes of little use. We point out that the spirit of such a change of measure is profoundly different from the change of measure argument for Large Deviations lower bounds, like the one we have used in Chap. 5 [cf. (5.19)]: as a matter of fact, our change of measure yields upper bounds.
3. *From correlation length to full system length: coarse graining estimates.* In order to go beyond annealed correlation length sizes we will look at the system as if it was made up of blocks of size k. We aim at showing that a visit to a block of size k essentially gives negative energetic reward, so that the *k-blocks coarse grained system* is delocalized. The passage to the coarse grained system is not straightforward: as a matter of fact we will just find upper bounds on the

partition function of the coarse grained system. It is also rather technical and we have preferred to move the main coarse grained estimates to the next chapter, i.e. Chap. 7.

6.2 The Basic Fractional Moment Estimates

The elementary, but crucial, observations that are at the basis of what we call *fractional moment method* are:

1. For every $\gamma \in (0,1]$, by Jensen inequality we have

$$\mathrm{F}_N(\beta,h) = \frac{1}{\gamma N}\mathbb{E}\left[\log Z_{N,\omega}^{\gamma}\right] \le \frac{1}{\gamma N}\log \mathbb{E}\left[Z_{N,\omega}^{\gamma}\right] \tag{6.2}$$

so that, in particular, $\sup_N \mathbb{E} Z_{N,\omega}^{\gamma} < \infty$ implies $\mathrm{F}(\beta,h)=0$. Note that for $\gamma=1$ this is the annealed bound.

2. For every $\gamma \in (0,1]$ and every $a_i \ge 0$

$$\left(\sum_i a_i\right)^{\gamma} \le \sum_i a_i^{\gamma}, \tag{6.3}$$

which is a simple consequence of $(1+x)^{\gamma} \le 1+x^{\gamma}$ for $x \ge 0$.

Remark 6.2. In view of what we have to show it is useful to replace h by $h-\log \mathrm{M}(\beta)$, which coincides with $h+h_c^a(\beta)$ since $\sum_n K(n)=1$, for all the rest of the chapter, so that

$$Z_{N,\omega} = Z_{N,\omega,\beta,h-\log \mathrm{M}(\beta)}, \tag{6.4}$$

till the end of the chapter.

A quick, but already profitable [31], way to exploit (6.2) and (6.3) is to apply it to the decomposition

$$Z_{N,\omega} = \sum_{n=1}^{N} \sum_{\substack{\ell \in \mathbb{Z}^{n+1}:\\ 0=\ell_0<\ell_1<\ldots<\ell_n=N}} \prod_{i=1}^{n} K(\ell_i-\ell_{i-1})\exp\left(\beta\omega_{\ell_i} - \log \mathrm{M}(\beta)+h\right), \tag{6.5}$$

so that

$$\begin{aligned}\mathbb{E}\left[Z_{N,\omega}^{\gamma}\right] &\le \sum_{n=1}^{N} \sum_{\substack{\ell \in \mathbb{Z}^{n+1}:\\ 0=\ell_0<\ell_1<\ldots<\ell_n=N}} \prod_{i=1}^{n} (K(\ell_i-\ell_{i-1}))^{\gamma}\mathbb{E}\exp\left(\gamma\beta\omega_1 - \gamma\log \mathrm{M}(\beta)+\gamma h\right)\\ &= \sum_{n=1}^{N} \exp\left(g_{\beta,\gamma} n\right) \sum_{\substack{\ell \in \mathbb{Z}^{n+1}:\\ 0=\ell_0<\ell_1<\ldots<\ell_n=N}} \prod_{i=1}^{n} K_{\gamma}(\ell_i-\ell_{i-1}),\end{aligned} \tag{6.6}$$

where we have assumed $(1+\alpha)\gamma > 1$, we have set $K_\gamma(n) := K(n)^\gamma/\Sigma_\gamma$, with

$$\Sigma_\gamma := \sum_{n=1}^{\infty} (K(n))^\gamma < \infty, \tag{6.7}$$

and

$$g_{\beta,\gamma} := \log \mathrm{M}(\beta\gamma) - \gamma \log \mathrm{M}(\beta) + \gamma h + \log \Sigma_\gamma. \tag{6.8}$$

Therefore if we introduce the (recurrent) $K_\gamma(\cdot)$-renewal $\widetilde{\tau}$ we have

$$\mathbb{E}\left[Z_{N,\omega}^\gamma\right] \le \mathbf{E}\left[\exp\left(g_{\beta,\gamma}|\widetilde{\tau}\cap(0,N]|\right); N \in \widetilde{\tau}\right], \tag{6.9}$$

so that $g_{\beta,\gamma} \le 0$ is a sufficient condition for $\mathrm{F}(\beta, h + \log \mathrm{M}(\beta)) = 0$. We sum up this with the following statement.

Proposition 6.3. *With Σ_γ as in* (6.7) *we have $h_c(\beta) \ge h_c^-(\beta)$ for every β, with*

$$h_c^-(\beta) := \max_{\gamma\in(1/(1+\alpha),1]} \frac{1}{\gamma} \log\left(\frac{\mathrm{M}(\beta)^\gamma}{\mathrm{M}(\beta\gamma)\Sigma_\gamma}\right) + h_c^a(\beta). \tag{6.10}$$

The strength of this statement lies of course in the fact that there are important cases in which the maximum of the expression in the right-hand side of (6.10) is achieved for a value of $\gamma < 1$ (if it is achieved for $\gamma = 1$, (6.10) reduces to the annealed bound). Notably, if

$$\lim_{\beta\to\infty} \frac{\mathrm{M}(\beta)^\gamma}{\mathrm{M}(\beta\gamma)} = +\infty, \tag{6.11}$$

a condition that is for example implied by $\log \mathrm{M}(\beta) \overset{\beta\to\infty}{\sim} c\beta^b$ (c and $b > 0$: $c = 1/2$ and $b = 2$ in the Gaussian case), then we have that

1. $h_c(\beta) > h_c^a(\beta) = 0$ for β sufficiently large.
2. $\lim_{\beta\to\infty} h_c(\beta) = +\infty$. In fact by choosing γ close to $1/(1+\alpha)$ one easily sees that, for every $\varepsilon > 0$, $h_c(\beta)$ is larger than $(1-\varepsilon)c\beta^b(\alpha/(1+\alpha))$ for β sufficiently large.

In words, if ω_1 is an unbounded random variable (with some weak condition on the tail decay) then for β sufficiently large, that is if there is enough disorder, quenched and annealed critical points differ (and they differ more and more when β gets larger) (Table 6.1).

Remark 6.4. One can of course extract more from (6.9). Notably that if $g_{\beta,\gamma} < 0$ then the right-hand side is just the partition function of a homogeneous model in the delocalized regime and therefore (cf. Theorem 2.7) there exists $c = c(g_{\beta,\gamma}, \gamma, K(\cdot))$

Table 6.1 A quantitative application of Proposition 6.3 with $\omega_1 \sim \mathcal{N}(0,1)$, $K(n) = c_K/n^{3/2}$ and $\sum_n K(n) = 1$. The quantity $\gamma_{\max}$ is the value of γ that achieves the maximum in the right-hand side of (6.10): $\gamma_{\max}$ approaches $2/3$ as β becomes large and $\gamma_{\max}$ is 1, and therefore $h_c^-(\beta) - h_c^a(\beta) = 0$, for $\beta < 2.5369\ldots$

β	$h_c^-(\beta) - h_c^a(\beta)$	$\gamma_{\max}$
2.7	0.0056	0.9741
3.0	0.0461	0.9333
4.0	0.4651	0.8405
6.0	2.5944	0.7530
10.0	11.6414	0.6985
20.0	59.4619	0.6744

such that

$$\mathbb{E}\left[Z_{N,\omega}^{\gamma}\right] \leq \frac{c}{N^{\gamma(1+\alpha)}}, \tag{6.12}$$

for every N.

6.3 The $\alpha > 1$ Case

6.3.1 A Different Look on Proposition 6.3

In order to go beyond the estimates in the previous section we are going to introduce suitable coarse graining procedures. We start by presenting a first, rather elementary, coarse graining that turns out to be performing well when $\alpha \geq 1$. In order to have an intuitive approach to it, let us quickly review the argument in the previous section by taking a slightly different viewpoint.

Let us apply (6.2) to the renewal identity

$$Z_{N,\omega} = \sum_{n=0}^{N-1} Z_{n,\omega} K(N-n) z_N, \quad \text{with } z_N := \exp(\beta\omega_N - \log \mathrm{M}(\beta) + h), \tag{6.13}$$

and, by taking the expectation, with the notation

$$A_n(\gamma, \beta, h) = A_n := \mathbb{E} Z^{\gamma}_{n,\omega,\beta,h+\log \mathrm{M}(\beta)}, \tag{6.14}$$

one gets to the *renewal inequality*

$$A_N \leq \mathbb{E}\left[z_1^{\gamma}\right] \sum_{n=0}^{N-1} A_n K(N-n)^{\gamma} = \sum_{n=1}^{N} A_{N-n} Q(n), \tag{6.15}$$

where $Q(n) := \mathbb{E}[z_1^{\gamma}] K(n)^{\gamma}$. Now the point is that (6.15) implies

$$A_N \leq \left(\sum_{n=1}^{\infty} Q(n)\right) \max_{n=0,1,\ldots,N-1} A_n, \tag{6.16}$$

so that, if $\sum_n Q(n) \leq 1$ we have $A_N \leq A_0 = 1$ for every N. Summing everything up

$$\mathbb{E}[z_1^\gamma] \sum_{n=1}^{\infty} K(n)^\gamma = \exp(\gamma h) \frac{\mathrm{M}(\gamma\beta)}{\mathrm{M}(\beta)^\gamma} \sum_{n=1}^{\infty} K(n)^\gamma \leq 1 \implies \sup_N A_N \leq 1 \tag{6.17}$$

and therefore once again to the result in Proposition 6.3.

By looking at (6.13) and (6.15) one gets a hint about the limits of the procedure: we are looking at one charge at a time. We are now going to present a method that looks at blocks of k charges.

6.3.2 A First Coarse Graining Procedure: Iterated Fractional Moment Estimates

Let us fix $k \in \mathbb{N}$: for every $N > k$ we can write the following renewal identity

$$Z_{N,\omega} = \sum_{n=k+1}^{N} Z_{N-n,\omega} \sum_{j=0}^{k} K(n-j)\, z_{N-j} Z_{j,\theta^{N-j}\omega}, \tag{6.18}$$

which is simply obtained by decomposing the partition function according to the value $N-n$ of the last point of τ before $N-k$ ($0 \leq N-n < N-k$ in the sum), and to the value $N-j$ of the first point of τ to the right of, or sitting on, $N-k$ (so that $N-k \leq N-j \leq N$). Of course $Z_{j,\theta^{N-j}\omega}$ has the same law as $Z_{j,\omega}$ and the three random variables $Z_{N-n,\omega}$, z_{N-j} and $Z_{j,\theta^{N-j}\omega}$ are independent for n and j in the whole range of summation (Fig. 6.1).

We now fix a $\gamma \in (0,1)$, apply (6.3) to (6.18) and take the expectation with respect to the charges to get for $N > k$

$$A_N \leq \mathbb{E}[z_1^\gamma] \sum_{n=k+1}^{N} A_{N-n} \sum_{j=0}^{k} K(n-j)^\gamma A_j. \tag{6.19}$$

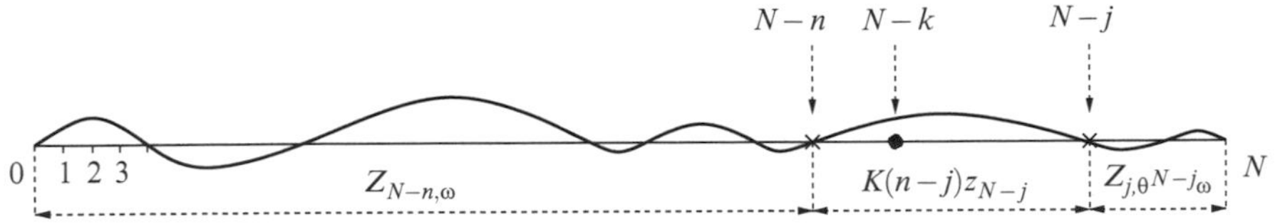

Fig. 6.1 The renewal identity (6.18) is obtained by fixing a value of k and summing over the values of the last contact (a cross) before $N-k$ (a site that is marked by a large dot) and the first contact (the second cross) after, and including, $N-k$. The two contacts are respectively $N-n$ and $N-j$

This is still a renewal inequality since it can be rewritten as

$$A_N \le \sum_{n=1}^{N} A_{N-n} Q_k(n), \tag{6.20}$$

with $Q_k(n) := \mathbb{E}[z_1^\gamma]\sum_{j=0}^{k} K(n-j)^\gamma A_j$ if $n > k$ and $Q_k(n) := 0$ for $n \le k$. In particular if for given β and h one can find $k \in \mathbb{N}$ and $\gamma \in (0,1)$ such that

$$\rho := \sum_n Q_k(n) = \mathbb{E}[z_1^\gamma] \sum_{n=k+1}^{\infty} \sum_{j=0}^{k} K(n-j)^\gamma A_j \le 1, \tag{6.21}$$

then one directly extracts from (6.20) that

$$A_N \le \rho \max\{A_0, \dots, A_{N-k-1}\}, \tag{6.22}$$

for $N > k$, which implies that $A_N \le \max\{A_0, \dots, A_k\}$ and hence $\mathrm{F}(\beta, h) = 0$.

We have therefore proven.

Proposition 6.5. *If, for given β and h, we can find $\gamma \in (0,1)$ and $k \in \mathbb{N}$ such that ρ, given in* (6.21)*, does not exceed one, then* $\mathrm{F}(\beta, h + \log \mathrm{M}(\beta)) = 0$.

Remark 6.6. Like in Remark 6.4 one can be sharper by exploiting the renewal structure in (6.20). The difference with Remark 6.4 is that in this case $N > k$. In order to put (6.20) into a more customary *renewal form* we set $\widetilde{A}_N := A_N \mathbf{1}_{N>k}$, so that

$$\widetilde{A}_N \le \sum_{n=1}^{N-k-1} \widetilde{A}_{N-n} Q_k(n) + P_k(N), \quad \text{with } P_k(N) = \sum_{n=0}^{k} A(n) Q_k(N-n), \tag{6.23}$$

and we observe that there exists $c > 0$ (depending on k, $K(\cdot)$ and γ, besides of course β and h) such that $P_k(N) \le c Q_k(N)$. Let us now consider the standard renewal equation for the $Q_k(n)$-renewal: $u_0 = 1$ (but of course one can choose an arbitrary $u_0 > 0$) and

$$u_N = \sum_{n=1}^{N} u_{N-n} Q_k(n) \overset{N \ge k}{=} \sum_{n=1}^{N-k-1} u_{N-n} Q_k(n) + u_0 Q_k(N), \tag{6.24}$$

where the first equality holds for $N = 1, 2, \dots$ and for the second one we have used that, since $Q_k(n) = 0$ up to $n = k$, we have $u_1 = u_2 = \dots = u_k = 0$. Once again if $\sum_n Q_k(n) < 1$, that is if $\rho < 1$, we are dealing with a renewal equation of a terminating process and therefore u_N behaves asymptotically like (a constant times) $Q_k(N)$ (by Theorem A.2). Comparing (6.23) and (6.24) one obtains that there exists a constant $C = C(K(\cdot), k, \gamma, h, \beta)$ such that

$$A_N \le \frac{C}{N^{(1+\alpha)\gamma}}. \tag{6.25}$$

6.3.3 Finite Volume Estimates: The Proof of Theorem 6.1 for $\alpha > 1$

Let assume $\sum_n K(n) = 1$. Let us start off by fixing γ:

$$\gamma := \frac{3+\alpha}{2+2\alpha} \quad \text{so that} \quad (1+\alpha)\gamma > 2. \tag{6.26}$$

It is clearly sufficient to show that $\mathrm{F}(\beta, \log \mathrm{M}(\beta) + \mathrm{c}\beta^2) = 0$ for $\beta \in (0, \beta_*]$ (so $Z_{N,\omega}$ stands for $Z_{N,\omega,\beta,\log \mathrm{M}(\beta)+\mathrm{c}\beta^2}$ from now till the end of the proof) and we are going to show this via Proposition 6.5, that has reduced the problem to a finite volume estimate. Notice in fact that estimating ρ, cf. (6.21), amounts to estimating only (a fractional moment of) $Z_{j,\omega}$, for $j \le k$. For this we need a new ingredient, but let us start with the preliminary observations:

- For $j = 0, \ldots, k$ we have $\sum_{n=k+1}^{\infty} K(n-j)^\gamma \le c_1 (k-j+1)^{1-(1+\alpha)\gamma}$, where $c_1 > 0$ depends only on $K(\cdot)$ (γ is a function of α, which is determined once $K(\cdot)$ is chosen).
- $\mathbb{E} z_1^\gamma \le c_2$ for every $\beta \in (0, \beta_*]$, with c_2 dependent only on β_*, γ and the law $\mathbb{P}$ of the charges (in view of the result we want to prove we can a priori assume c is smaller than a fixed constant, for example $\mathrm{c} \le 1$, so that $h \in (0, \beta_*^2])$.

Therefore $\rho \le 1$ if

$$\sum_{j=0}^{k} \frac{A_j}{(k-j+1)^{(1+\alpha)\gamma-1}} \le \frac{1}{c_1 c_2} =: \varepsilon, \tag{6.27}$$

and therefore $\varepsilon = \varepsilon(K(\cdot), \beta_*, \mathbb{P})$. We will now show that the expression in the left-hand side of (6.27) can be made small choosing k suitably (large).

To this purpose let us observe that we know of course (Jensen inequality) that

$$\begin{aligned} A_j \le (\mathbb{E} Z_{j,\omega})^\gamma &= \left(\mathbf{E} \left[\exp\left(\mathrm{c}\beta^2 \sum_{n=1}^{j} \delta_n \right) ; j \in \tau \right] \right)^\gamma \\ &= \exp\left(\gamma \mathrm{F}(0, \mathrm{c}\beta^2) j\right) \mathbf{P}\left(j \in \widetilde{\tau}^{(\mathrm{c}\beta^2)} \right)^\gamma, \end{aligned} \tag{6.28}$$

where we have use the notation of Remark 2.3 for the renewal pinned down by the homogeneous potential $\mathrm{c}\beta^2$. Since $\mathrm{c}\beta^2 > 0$, by the Renewal Theorem the probability that $j \in \widetilde{\tau}^{(\mathrm{c}\beta^2)}$ is bounded below by a positive constant (even if $\mathrm{c}\beta^2$ were zero!), so this term cannot be of much help and we simply bound it above by one. On the other hand the exponentially growing term stays bounded for j up to the correlation length of the annealed system (cf. Sect. 2.4 of Chap. 2): we therefore choose

$$k = k(\beta, \mathtt{c}) := \left\lfloor \frac{1}{\mathtt{F}(0, \mathtt{c}\beta^2)} \right\rfloor . \tag{6.29}$$

Note that, because of this choice of k and (6.28) we have that $A_j \le \exp(\gamma)$ for every $j \le k$. Therefore, because of (6.26), the expression in the left-hand side of (6.27) is bounded even for k large: since we want to get estimates that hold for every $\beta \in (0, \beta_*]$, by k large we mean, here and below, $\mathtt{c}$ small and how small depends only on $K(\cdot)$, β_* and $\mathbb{P}$. This is not yet what we want (the leftmost side of (6.27) has to be made smaller than a suitable small constant and for now we just know that it is bounded), but a more attentive analysis shows that one has

$$\sum_{j=0}^{k-R} \frac{A_j}{(k-j+1)^{(1+\alpha)\gamma-1}} \le \exp(\gamma) \sum_{j>R} j^{-(1+\alpha)\gamma+1} \le \frac{\varepsilon}{2}, \tag{6.30}$$

for any $k > R$ and R chosen sufficiently large (how large depends only on only on α and ε, that is only on $K(\cdot)$ and β_*). This has been achieved by using (6.28). We are therefore left with showing that

$$\sum_{j=k-R}^{k} \frac{A_j}{(k-j+1)^{(1+\alpha)\gamma-1}} \le \frac{\varepsilon}{2}. \tag{6.31}$$

For this we set [recall (6.14)]

$$\widehat{A} := \limsup_{\mathtt{c}\searrow 0} \sup_{\beta\in(0,\beta_*]} \max_{j=k-R+1,\dots,k} A_j(\gamma, \beta, \mathtt{c}\beta^2). \tag{6.32}$$

If we are able to show that

$$\widehat{A} \sum_{i=1}^{R} i^{-((1+\alpha)\gamma-1)} \le \frac{\varepsilon}{3}, \tag{6.33}$$

then (6.27) would be established for $\mathtt{c}$ small. Of course in the upper limit for i in (6.33) one can replace R with ∞ obtaining thus a more stringent condition (but, in the end, equivalent, since we are not tracking the constants). We stress that R has been chosen at this stage of the proof and will be kept fixed.

In order to go beyond (6.28), and prove (6.33), the new idea is a tilting procedure, that reduces to a shift when the charges are Gaussian variables. The idea is based on the following consequence of Hölder inequality

$$A_j = \widetilde{\mathbb{E}}\left[(Z_{j,\omega})^\gamma \frac{\mathrm{d}\mathbb{P}}{\mathrm{d}\widetilde{\mathbb{P}}}(\omega) \right] \le \widetilde{\mathbb{E}}\left[\left(\frac{\mathrm{d}\mathbb{P}}{\mathrm{d}\widetilde{\mathbb{P}}}(\omega) \right)^{1/(1-\gamma)} \right]^{1-\gamma} \widetilde{\mathbb{E}}\left[Z_{j,\omega} \right]^\gamma, \tag{6.34}$$

where $\widetilde{\mathbb{P}}$ is a probability equivalent to $\mathbb{P}$ (i.e. $\mathbb{P}$ and $\widetilde{\mathbb{P}}$ are mutually absolutely continuous).

The case of Gaussian charges. For Gaussian charges we choose $\widetilde{\mathbb{P}}$ to be the law of the sequence

$$\omega_1 - \sqrt{c\beta^2}, \omega_2 - \sqrt{c\beta^2}, \ldots, \omega_k - \sqrt{c\beta^2}, \omega_{k+1}, \omega_{k+2}, \ldots \tag{6.35}$$

which is a sequence of independent (non identically distributed) variables. By using $c\beta^2 k \leq c\beta^2/\mathrm{F}(0, c\beta^2)$ and the fact that $\lim_{\delta \searrow 0} \delta/\mathrm{F}(0,\delta) =: c_{\mathrm{F}}(K(\cdot)) > 0$ since $\alpha > 1$, cf. Theorem 2.10, one then readily obtains

$$\widetilde{\mathbb{E}}\left[\left(\frac{d\mathbb{P}}{d\widetilde{\mathbb{P}}}(\omega)\right)^{1/(1-\gamma)}\right]^{1-\gamma} = \\ \exp\left(\frac{\gamma}{2(1-\gamma)} c\beta^2 k\right) \leq \exp\left(\frac{c_{\mathrm{F}}\gamma}{1-\gamma}\right) = \exp\left(c_{\mathrm{F}}\frac{3+\alpha}{\alpha-1}\right), \tag{6.36}$$

where the inequality holds for c small.

Let us now turn our attention to $\widetilde{\mathbb{E}} Z_{j,\omega}$ which, for $j \leq k$, is simply $\mathbb{E}Z_{j,\omega-\sqrt{c\beta^2}}$. But this is just the partition function of a homogeneous model with negative pinning potential if we choose c small, namely

$$\begin{aligned} \widetilde{\mathbb{E}} Z_{j,\omega} &= \mathbf{E}\left[\exp\left(-\beta^2(\sqrt{c}-c)\,|\tau \cap (0,j]|\right); j \in \tau\right] \\ &= \mathbf{E}\left[\exp\left(-\left(\frac{1}{\sqrt{c}}-1\right)(c\beta^2 k)\left(\frac{j}{k}\right)\frac{|\tau \cap (0,j]|}{j}\right); j \in \tau\right]. \end{aligned} \tag{6.37}$$

But $\lim_{j\to\infty} |\tau \cap (0,j]|/j = 1/\mathbf{E}[\tau_1]$ $\mathbf{P}$-a.s. and this readily implies that, uniformly in $\beta \in (0, \beta_*]$, $\sup_{j=k-R,\ldots,k} \widetilde{\mathbb{E}} Z_{j,\omega}$ is bounded by a constant that can be chosen arbitrarily small, provided one chooses c sufficiently small (so $c^{-1/2} - 1$ is large). In detail: by choosing c sufficiently small, we can make sure that

$$\inf_{\beta \in (0,\beta_*]} k(\beta, c) \geq 2R \quad \text{and} \quad \inf_{\beta \in (0,\beta_*]} c\beta^2 k \geq \frac{2}{3}\lim_{a \searrow 0}\frac{a}{\mathrm{F}(0,a)} = \frac{2}{3} c_{\mathrm{F}}(K(\cdot)) > 0, \tag{6.38}$$

so that by (6.34), (6.36) and (6.37) we see that

$$\widehat{A} \leq \exp\left(c_{\mathrm{F}}\frac{3+\alpha}{\alpha-1}\right) \limsup_{c \searrow 0} \max_{j=k-R,\ldots,k} \mathbf{E}\left[\exp\left(-\frac{c_{\mathrm{F}}}{3}\left(\frac{1}{\sqrt{c}}-1\right)\frac{|\tau \cap (0,j]|}{j}\right)\right]^{\gamma}. \tag{6.39}$$

Therefore we are done if we can show that the right-hand side is smaller than $\varepsilon/(3\sum_i i^{1-(1+\alpha)\gamma})$. For this it suffices to choose a value c_0 sufficiently small to guarantee

$$\exp\left(c_{\mathrm{F}}\frac{3+\alpha}{\alpha-1}\right)\exp\left(-\frac{c_{\mathrm{F}}\gamma}{4}\left(\frac{1}{\sqrt{\mathrm{c}_0}}-1\right)\frac{1}{\mathbf{E}[\tau_1]}\right)\le\frac{\varepsilon}{6\sum_i i^{1-(1+\alpha)\gamma}}. \tag{6.40}$$

If we now replace $1/\sqrt{\mathrm{c}}$ with $1/\sqrt{\mathrm{c}_0}$ in the right-hand side of (6.39) we obtain an upper bound and the result follows by the Renewal Theorem. This completes the proof of Theorem 6.1, for $\alpha > 1$, in the case of Gaussian charges.

The case of general charges. The proof in the general case, i.e. for charges that are not necessarily Gaussian, is very similar. Shifting the charges, cf. (6.35), carries over to other continuous random variables with positive density, but in general we need to resort to *tilting*. The new measure is therefore chosen, for $n \in \mathbb{N}$ and $\lambda \in \mathbb{R}$, to be

$$\frac{\mathrm{d}\widetilde{\mathbb{P}}_{n,\lambda}}{\mathrm{d}\mathbb{P}}(\omega)=\frac{1}{\mathrm{M}(-\lambda)^n}\exp\left(-\lambda\sum_{i=1}^n\omega_i\right). \tag{6.41}$$

Like before we choose $n = k$, k is still the one in (6.29), and $\lambda = \sqrt{\mathrm{c}\beta^2}$: let us assume like before that $\mathrm{c} \le 1$ so that $\lambda \le \beta$. We set $\widetilde{\mathbb{P}} := \widetilde{\mathbb{P}}_{k,\sqrt{\mathrm{c}\beta^2}}$ and we go back to (6.34): we have to estimate two terms. For the first we observe that

$$\widetilde{\mathbb{E}}\left[\left(\frac{\mathrm{d}\mathbb{P}}{\mathrm{d}\widetilde{\mathbb{P}}}\right)^{1/(1-\gamma)}\right]^{1-\gamma}$$
$$=\left(\mathrm{M}\left(-\sqrt{\mathrm{c}\beta^2}\right)^{\gamma}\mathrm{M}\left(\sqrt{\mathrm{c}\beta^2}\gamma/(1-\gamma)\right)^{1-\gamma}\right)^k\le\exp\left(\frac{C_{\mathrm{M}}\gamma}{1-\gamma}\right) \tag{6.42}$$

where $2C_{\mathrm{M}} := \max_{|t|\le 1}(\log \mathrm{M}(t))''$ and we have assumed c smaller than $\beta_*\gamma/(1-\gamma)$ (so that the arguments of $\mathrm{M}(\cdot)$ in the middle term in (6.42) are in $[-1,1]$).

For what concerns the second term:

$$\widetilde{\mathbb{E}}[Z_{j,\omega}]=\mathbf{E}\left[\left(\exp\left(\left(\mathrm{c}-\sqrt{\mathrm{c}}\right)\beta^2\right)\frac{\mathrm{M}(\beta-\beta\sqrt{\mathrm{c}})}{\mathrm{M}(\beta)\mathrm{M}(-\beta\sqrt{\mathrm{c}})}\right)^{|\tau\cap\{1,\dots,j\}|};j\in\tau\right]. \tag{6.43}$$

This formula can be made more exploitable by observing that, if we set C_{M,β_*} equal to $\min_{|t|\le\beta_*}(\log \mathrm{M}(t))''(>0)$, we have for $0<\lambda\le\beta\le\beta_*$ and

$$\frac{\mathrm{M}(\beta-\lambda)}{\mathrm{M}(\beta)\mathrm{M}(-\lambda)}=\exp\left(-\int_0^\beta \mathrm{d}x\int_{-\lambda}^0\mathrm{d}y\,\frac{\mathrm{d}^2}{\mathrm{d}t^2}\log \mathrm{M}(t)\bigg|_{t=x+y}\right)\le e^{-C_{\mathrm{M},\beta_*}\beta\lambda}. \tag{6.44}$$

The proof is then completed precisely like in the Gaussian case. □

6.4 The $\alpha = 1$ Case

In the $\alpha > 1$ case we have used that $(1+\alpha)\gamma > 2$. This is no longer possible if $\alpha = 1$ and one needs to be a bit more careful in estimating ρ [cf. (6.21)]. Essentially three extra ingredients are going to be used in showing that ρ can be made small by choosing $\mathtt{c}$ small for $\alpha = 1$:

- A suitable choice of $\gamma = \gamma(k)$, with $\gamma(k) \nearrow 1$ as $k \to \infty$
- Exploiting the pinning term that is present in the partition function: this time $\lim_N \mathbf{P}(N \in \tau) = 0$
- Compared to (6.30) and (6.31), a different splitting of the summation defining ρ (and the rough bounds obtained by replacing finite sums by the series in this new case would lead to divergences!)

Proof of Theorem 6.1, $\alpha = 1$. The $\alpha = 1$ case presents all the troubles related to the fact that it makes appear slowly varying functions (as a matter of fact, only logarithms, but that is annoying enough). So things get somewhat lengthy and involved. Not to make the presentation too heavy and not to detour the reader from the main issues we will give only the essential ingredients of the proof, but we will be rather sketchy with the intermediate steps. In any case the reader should take what follows as the guided solution to a difficult, or, at least, lengthy, exercise: the advise for the interested reader is to come back to $\alpha = 1$ after having looked at the detailed treatment of $\alpha \in (1/2, 1)$. We point out also that, were one shooting for the weaker statement that for any $\varepsilon > 0$ there exists $c_\varepsilon > 0$ such that $h_c(\beta) - h_c^a(\beta) \geq c_\varepsilon \beta^{2-\varepsilon}$ for every $\beta \leq \beta_*$, the proof would be definitely quicker.

It is of course sufficient to establish delocalization for $h = \mathtt{c}\beta^2/(\log(1+1/\beta))^2$ (as before, we are looking at the system with partition function $Z_{N,\omega,\beta,-\log \mathrm{M}(\beta)+h}$). In accord with Theorem 2.10, we choose $k = k(\beta, \mathtt{c})$ equal to (the integer part of) $(\log(1+1/\beta))^3/(\mathtt{c}\beta^2)$. Moreover we set $\gamma = \gamma(k) = 1 - (1/\log k)$. We bound ρ from above by bounding (6.27): we split the sum in two terms, T_1 and T_2, according to whether j is smaller (T_1) or larger (T_2) than $y(\mathtt{c})k$, with $\lim_{\mathtt{c}\searrow 0} y(\mathtt{c})k(\beta,\mathtt{c}) = 0$, but $\lim_{\mathtt{c}\searrow 0} y(\mathtt{c}) = 0$ (a definite choice is made below).

For T_1 it suffices bound A_j by Jensen inequality, like we did for (6.30), and to observe that that $(k-j+1)^{-2\gamma+1} = (k-j+1)^{-1}(k-j+1)^{2/\log k}$ so that, for j is smaller than yk, the factor $(k-j+1)^{2/\log k}$ is bounded by a constant. This yields $T_1 = O(y)$, that is T_1 can be made small by choosing $\mathtt{c}$ small.

For T_2 we have to be more careful:

- We apply a change of measure argument, like for (6.31), by shifting like in (6.35) (or tilting, if non Gaussian), but by $\sqrt{\mathtt{c}}\beta/(\log(1+1/\beta))^2$ and not by $\sqrt{\mathtt{c}}\beta$. We still use (6.34) and a computation shows that the first factor in the right-hand side of (6.34) is bounded by a constant with the choice of parameters we have made. We turn then to the second factor that is an annealed partition function to the power γ.

- In order to bound this annealed partition we first exploit the fact that $j \in \tau$ and we recall that [cf. (A.10)] $\mathbf{P}(j \in \tau) \sim 1/(\mathtt{c}_K \log j)$ for j large, and, with our choice of $\gamma = \gamma(k)$, we have also $\mathbf{P}(j \in \tau)^{\gamma(k)} \sim 1/(\mathtt{c}_K \log j)$ for j large (of course $j \in (yk, k]$). In order to take advantage of this term we need to condition with respect to $j \in \tau$ (like, for example in (6.51) below). This conditioning is unpleasant to deal with directly, but it can be removed by using Lemma A.5.
- What one is left with is evaluating for j large (rather, one should say k large) an expression of the type

$$\mathbf{E}\left[\exp\left(c_1 y(\mathtt{c}) \frac{(\mathtt{c} - \sqrt{\mathtt{c}})}{\mathtt{c}} \frac{|\tau \cap (0, j]|}{j/\log j}\right)\right], \tag{6.45}$$

  with c_1 a positive constant that can be explicitly evaluated. At this point we require $y(\mathtt{c})$ not to go to zero too fast when $\mathtt{c}$ goes to zero: we need in fact $y(\mathtt{c})(\mathtt{c} - \sqrt{\mathtt{c}})/\mathtt{c}$ to be large (and negative!) for $\mathtt{c}$ small (so, for example, choose $y(\mathtt{c}) = \mathtt{c}^{1/4}$).
- Finally we observe that the random variable $|\tau \cap (0, j]|/(j/\log j)$ converges in probability as $j \to \infty$ to a positive constant: this can be proven by hand by arguing like in the proof of Proposition A.6. The argument is actually easier that for the case $\alpha \in (0, 1)$ because τ_1 for $\alpha = 1$ is not in the domain of attraction of a stable law, but we have rather the phenomenon of *relative stability* (see [7, Sect. 8 of Chap. 8]), in the sense that τ_n, suitably normalized converges in law, but to a degenerate non zero variable. For what interests us one verifies (either directly or by applying [7, Theorem 8.8.1] that $\tau_n/(\mathtt{c}_K n \log n)$ converges to 1. Therefore the expression in (6.45) can be made small by choosing $\mathtt{c}$ small.

The ensemble of the elements we have provided leads to the fact ρ can be made arbitrarily small by choosing $\mathtt{c}$ small. We just need to make it smaller than one, so we are done. □

6.5 The $\alpha \in (1/2, 1)$ Case

The starting point is, like for the previous section, that in the $\alpha > 1$ case we have used in an important way that $(1 + \alpha)\gamma > 2$, see for example (6.33). Like for $\alpha = 1$, if $\alpha < 1$ the very same strategy is no longer possible, but when $\alpha > 1$ we have not gained anything from the *last site pinning* that is present in $Z_{n,\omega}$: see for example (6.28). Now we want (and we need, if we want to get precise estimates, see Sect. 6.7) to keep track of this extra gain.

In this new context the *iterated fractional estimates* of the previous sections still work, but they yield a result that is a bit weaker than what is claimed in Theorem 6.1. Moreover the iterated fractional estimates are useless for the $\alpha = 1/2$ case and therefore we resort to a more involved coarse graining procedure.

The strategy of the proof is the one outlined in Sect. 6.1, but let us precise it a bit before going into the details (since we are giving for granted the use of fractional moments the steps now shrink to two). What we are going to do is:

1. Estimate $\mathbb{E}Z^{\gamma}_{j,\omega}$ for j up to the correlation length k of the pure system: such an estimate is obtained by a change of measure argument, at least for j of the order of k (in the $\alpha > 1$ case it was sufficient to use such an argument only for j's that differ from k only of a constant). Smaller values of j can be treated in a rougher way;
2. Upgrade the estimates up to the correlation length to estimates for arbitrary sizes via a coarse graining procedure.

The second step – the coarse graining – is treated in Chap. 7: Proposition 7.1 is giving us two sufficient conditions for $\textsc{f}(\beta,h) = 0$, and these conditions are precisely conditions *up to size k*. Proposition 7.1 has a rather technical appearance, also because it mixes a bit the steps (1) and (2) above, even if it is essentially about step (2). We will therefore start by proving a result that we will not use, but that contains the essence and all the technical difficulties of what we need, while being much more intuitive to grasp than the hypotheses of Proposition 7.1 (which we will verify right after).

Throughout this section we set

$$\gamma := \frac{2}{2+\alpha} \in (0,1)\,, \tag{6.46}$$

so that $(1+\alpha)\gamma > 1$, and

$$k = k(\beta,\mathtt{c}) := 2\left\lfloor \frac{1}{\textsc{f}\left(0, \mathtt{c}\beta^{\frac{2\alpha}{2\alpha-1}}\right)} \right\rfloor \overset{\mathtt{c}\searrow 0}{\sim} \frac{c_1}{\mathtt{c}^{\frac{1}{\alpha}}\beta^{\frac{2}{2\alpha-1}}}\,, \tag{6.47}$$

with $c_1 = c_1(K(\cdot))$ a positive constant (given in Theorem 2.10). The definition of k is consistent with the fact that we want to prove delocalization for h up to about $\beta^{2\alpha/(2\alpha-1)}$ and the pure correlation length diverges like the reciprocal of the (pure) free energy, that is, like $h^{-1/\alpha}$.

6.5.1 Bounds for Correlation Length Size Systems

Here is what we want to prove.

Proposition 6.7. *Choose a* $K(\cdot)$ *that satisfies* (2.30) *with* $\alpha \in (1/2,1)$, $\mathbb{P}$ *as in Hypothesis 3.1 and fix* $\beta_* > 0$. *For every* $\eta > 0$ *we can find* $\mathtt{c} > 0$ *such that*

$$\mathbb{E}Z^{\gamma}_{k,\omega} \le \eta\, \mathbf{P}(k \in \tau)^{\gamma}\,, \tag{6.48}$$

for every $h \le \mathtt{c}\beta^{2\alpha/(2\alpha-1)}$ *and every* $\beta \in (0,\beta_*]$.

Proof. We apply the change of measure procedure, that is (6.34), and like in the preceding section the new measure that we are going to choose is just a tilted version of the previous one. Once again, tilting reduces to shifting in the Gaussian context, and we present that first. We fix $h = \mathtt{c}\beta^{2\alpha/(2\alpha-1)}$ since this directly implies the full result.

The case of Gaussian charges. We set $\widetilde{\mathbb{P}}$ to be the law of

$$\omega_1 - \frac{1}{2\sqrt{k}}, \omega_2 - \frac{1}{2\sqrt{k}}, \ldots, \omega_k - \frac{1}{2\sqrt{k}}, \omega_{k+1}, \omega_{k+2}, \ldots. \tag{6.49}$$

This is absolutely analogous to (6.35), even quantitatively, because $1/\sqrt{k}$ is about $\mathtt{c}^{1/(2\alpha)}\beta^{1/(2\alpha-1)}$, which for $\alpha = 1$ matches with (6.35) (of course the choice of the constant pre-factors are just to get more compact expressions later on). We have:

$$\widetilde{\mathbb{E}}\left[\left(\frac{\mathrm{d}\mathbb{P}}{\mathrm{d}\widetilde{\mathbb{P}}}(\omega)\right)^{1/(1-\gamma)}\right]^{1-\gamma} = \exp\left(\frac{\gamma}{2(1-\gamma)}\left(\frac{1}{2\sqrt{k}}\right)^2 k\right) = \exp\left(\frac{1}{4\alpha}\right) \le 2. \tag{6.50}$$

On the other hand

$$\widetilde{\mathbb{E}}Z_{k,\omega} = \mathbf{E}\left[\exp\left(-\left(\frac{\beta}{2\sqrt{k}} - h\right)|\tau \cap (0,k]|\right)\Big| k \in \tau\right]\mathbf{P}(k \in \tau), \tag{6.51}$$

and notice that

$$\frac{\beta}{2\sqrt{k}} - h = \frac{\beta}{2\sqrt{k}} - \mathtt{c}\beta^{2\alpha/(2\alpha-1)} \ge c_2\left(\frac{k}{2}\right)^{-\alpha}\mathtt{c}^{-\frac{2\alpha-1}{2\alpha}}, \tag{6.52}$$

where the inequality is a consequence of (6.47) and holds for $\mathtt{c}$ sufficiently small (and $c_2 = c_2(K(\cdot)) > 0$). Let us go back to (6.51) and let us apply Lemma A.5 after having used $|\tau \cap (0,k]| \le |\tau \cap (0,k/2]|$:

$$\mathbf{E}\left[\exp\left(-\left(\frac{\beta}{2\sqrt{k}} - h\right)|\tau \cap (0,k]|\right)\Big| k \in \tau\right] \le \\ C_{bc}\,\mathbf{E}\left[\exp\left(-c_2\mathtt{c}^{-\frac{2\alpha-1}{2\alpha}}\frac{|\tau \cap (0,k/2]|}{(k/2)^\alpha}\right)\right]. \tag{6.53}$$

But $|\tau \cap (0,n]|/n^\alpha$ converges in law to the random variable Y_α (Proposition A.6: note that $\mathbf{P}(Y_\alpha > 0) = 1$), so that if $\mathtt{c}$ is sufficiently small (depending on $K(\cdot)$, as usual, but this time also on η, that is chosen in the statement) we have

$$\mathbf{E}\left[\exp\left(-c_2\mathtt{c}^{-\frac{2\alpha-1}{2\alpha}}Y_\alpha\right)\right] \le \frac{(\eta/2)^{1/\gamma}}{2C_{bc}}. \tag{6.54}$$

But this implies, going back to (6.51), that for c sufficienly small

$$\widetilde{\mathbb{E}} Z_{k,\omega} \le \left(\frac{\eta}{2}\right)^{1/\gamma} \mathbf{P}(k \in \tau) . \tag{6.55}$$

By injecting this estimate, together with (6.50), into (6.34) we obtain (6.48) and the proof of Proposition 6.7 is complete in the case of Gaussian charges.

The case of general charges. The tools to generalize the proof are (6.41) and (6.44). Once these formulas are applied one is back to expressions that differ from the Gaussian ones only for constants that depend on the law of ω_1. □

6.5.2 *Proof of Theorem 6.1, Case* $\alpha \in (1/2,1)$

We are going to verify the three conditions of Proposition 7.1. The correlation length is chosen like in the proof of Proposition 6.7, cf. (6.47), as well as h (still equal to $\mathrm{c}\beta^{2\alpha/(2\alpha-1)}$). We choose $\widetilde{\mathbb{P}}_k$ in Proposition 7.1 to be equal to $\widetilde{\mathbb{P}}$ given either in (6.49) (for the Gaussian case) or in (6.41) for the general case. Hypothesis (1) of Proposition 7.1 is therefore just (6.50). For what concerns Hypotheses (2) and (3) we observe that for $d, f \in \{1,2,\dots k\}$ and $d \le f$ we have

$$\widetilde{\mathbb{E}}_k \mathbf{E}\left[\exp\left(\sum_{n=d}^{f} (\beta\omega_n - \log \mathrm{M}(\beta) + h)\delta_n \right) \delta_f \,\middle|\, d \in \tau \right] = \\ \mathbf{E}\left[\exp(-a(\beta,\mathrm{c})|\tau \cap [d,f]|) \,\middle|\, d,f \in \tau \right] \mathbf{P}(f-d \in \tau), \tag{6.56}$$

where $a(\beta,\mathrm{c})$ can be written out explicitly without much effort, for example in the Gaussian case $a(\beta,\mathrm{c}) = -h + \beta/(2\sqrt{k})$ [see (6.51)] and what is important for us is that, for c sufficiently small we have $a(\beta,\mathrm{c}) \ge c_2 \mathrm{c}^{1/(2\alpha)} \beta^{2\alpha/(2\alpha-1)}$ for a suitable constant $c_2 = c_2(\beta_*, K(\cdot), \mathbb{P})$. The fact that $a \ge 0$ (for c sufficiently small) suffices to verify Hypothesis (3) and the 2 in the right-hand side of (7.7) can even be replaced by one. For Hypothesis (2) let us instead observe that, by (6.56), it suffices to verify that

$$\mathbf{E}\left[\exp(-a(\beta,\mathrm{c})|\tau \cap (0, f-d+1]|) \,\middle|\, f-d+1 \in \tau \right] \le \eta , \tag{6.57}$$

holds for c sufficiently small and $f - d > \varepsilon k$. But this is obtained by proceeding like in (6.53):

$$\begin{aligned} &\mathbf{E}\left[\exp(-a(\beta,\mathrm{c})|\tau \cap (0, f-d+1]|) \,\middle|\, f-d+1 \in \tau \right] \\ &\qquad \le C_{bc} \mathbf{E}\left[\exp(-a(\beta,\mathrm{c})\, |\tau \cap (0,(f-d+1)/2]|) \right] \\ &\qquad \le C_{bc} \mathbf{E}\left[\exp\left(-c_3 \varepsilon^{\alpha} \mathrm{c}^{(1-2\alpha)/2\alpha} \frac{|\tau \cap (0, \varepsilon k/2]|}{(\varepsilon k/2)^{\alpha}} \right) \right], \end{aligned} \tag{6.58}$$

where, in the last step, we have used the asymptotic behavior of k, cf. (6.29), so that, uniformly in $\beta \in (0, \beta_*]$, $a(\beta, \mathtt{c})(\varepsilon k/2)$ behaves like $c_2 \varepsilon^\alpha c_1^\alpha 2^{-\alpha} \mathtt{c}^{(1-2\alpha)/(2\alpha)}$ for $\mathtt{c}$ small (therefore we can take for example $c_3 = c_2 c_1^\alpha / 3$). We are therefore in the very same situation as for the right-hand side of (6.53): the only change is in the presence of ε in the prefactor and in the length of the interval, but the argument that follows (6.53) still applies with the straightforward change that now $\mathtt{c}$ will have to be taken smaller than a constant that depends on $K(\cdot)$, $\mathbb{P}$ (when the charges are not Gaussian), η and ε. □

6.6 The $\alpha = 1/2$ Case

Like the previous one, also this section is split into two parts: first we prove a result that we will not use in the proof of Theorem 6.1 ($\alpha = 1/2$), but it contains the essence of the argument that, in the second part, will lead to the proof. And like in the previous section we fix a value of γ once for all the section: $\gamma = 4/5$, which is precisely the choice in (6.46), so that $(1+\alpha)\gamma = 6/5 > 1$. Moreover the natural correlation length of the pure model is, as usual, $1/\mathrm{F}(0,h) \overset{h \searrow 0}{\sim} const.h^{-2}$ as $h \searrow 0$. In partial disagreement with this we are going to set

$$k = k(\beta, \mathtt{c}) := 2\left\lfloor \frac{1}{2h} \right\rfloor \overset{h \searrow 0}{\sim} \frac{1}{h}, \tag{6.59}$$

but we will still think of k as a correlation length for the system. We are allowed to do that because we are aiming at showing that when $h = \exp(-1/(\mathtt{c}\beta^4))$ (this gives the explicit expression that we are going to use for $k(\beta, \mathtt{c})$ in the proof) the free energy is zero: the *more natural* $1/h^2$ choice for the correlation length amounts to replacing $\mathtt{c}$ with $2\mathtt{c}$, so the two choices are equivalent because we are not tracking precisely the value of $\mathtt{c}$. On the other hand the choice (6.59) allows getting rid immediately of h in the partition function for systems of size $O(k)$, at the little expense of a multiplicative factor.

6.6.1 *Estimates up to the (Annealed) Correlation Length: Gaussian Case*

Here is the result that is analogous to Proposition 6.7 (we just give it for Gaussian charges: the case of general charges is treated when proving Theorem 6.1).

Proposition 6.8. *Choose $\omega_1 \sim \mathcal{N}(0,1)$. Fix $K(\cdot)$ satisfying (2.30) and $\alpha = 1/2$, and set $h = \exp(-\mathtt{c}\beta^{-4})$ and k as in (6.59). Then for every $\eta > 0$ there exists $\mathtt{c} > 0$ such that*

$$\mathbb{E}\left[Z_{k,\omega}^\gamma\right] \le \eta\, (\mathbf{P}(k \in \tau))^\gamma \tag{6.60}$$

for every $\beta \in (0, \beta_*]$.

Proof. As before, the first (and key) step for our estimates is to observe that if we introduce a probability $\widetilde{\mathbb{P}}$ which is equivalent to $\mathbb{P}$ we have

$$\mathbb{E}\left[Z_{k,\omega}^{\gamma}\right] \leq \widetilde{\mathbb{E}}\left[\left(\frac{\mathrm{d}\mathbb{P}}{\mathrm{d}\widetilde{\mathbb{P}}}(\omega)\right)^{\frac{1}{1-\gamma}}\right]^{1-\gamma} \widetilde{\mathbb{E}}\left[Z_{k,\omega}\right]^{\gamma}, \tag{6.61}$$

where we have used the Hölder inequality, with $p = 1/\gamma$. The quantity $(\mathrm{d}\mathbb{P}/\mathrm{d}\widetilde{\mathbb{P}})(\omega)$ is the density of the old measure with respect to the new one and, of course, we can restrict our attention to $\omega_1, \dots, \omega_k$ since the values of the charges $\omega_{k+1}, \omega_{k+2}, \dots$ do not contribute to $Z_{k,\omega}$. We choose $\widetilde{\mathbb{P}}$ to be a centered Gaussian measure too, but unlike $\mathbb{P}$ under which $\omega_1, \dots, \omega_k$ are IID, under $\widetilde{\mathbb{P}}$ the variables are correlated:

$$\widetilde{\mathbb{E}}\left[\omega_i \omega_j\right] = \begin{cases} 1 & \text{if } i = j, \\ -H_{i,j} & \text{if } i \neq j. \end{cases} \tag{6.62}$$

H is of course a symmetric matrix: we choose it to be traceless with $H_{i,j} \geq 0$. We actually make the definite choice (for $i \neq j$)

$$H_{i,j} = \frac{1-\gamma}{3\sqrt{|i-j| k \log k}}. \tag{6.63}$$

Since $I - H$ is a covariance matrix (and since we want to work with probability densities), it has to be positive definite: this is the case, at least if k is not too small. To see this it is convenient to introduce the Hilbert–Schmidt norm $\|\cdot\|$

$$\|H\|^2 := \sum_{i,j} H_{i,j}^2 = \sum_{j=1}^{k} \lambda_j(H)^2 \tag{6.64}$$

where $\lambda_1(H), \dots, \lambda_k(H)$ are the eigenvalues of H, so that in particular $\max_j |\lambda_j(H)|$ is bounded above by $\|H\|$. We have

$$\sum_{i,j} H_{i,j}^2 = \frac{2(1-\gamma)^2}{9k \log k} \sum_{i=1}^{k-1} \sum_{j=1}^{k-i} \frac{1}{j} \overset{k \to \infty}{\sim} \frac{2(1-\gamma)^2}{9}, \tag{6.65}$$

so that for k sufficiently large

$$\|H\| \leq \frac{1-\gamma}{2}, \tag{6.66}$$

which implies that the spectrum of $I - H$ is bounded below by $(1+\gamma)/2 = 9/10$, so $I - H$ is positive definite. We now make the first factor in the right-hand side of (6.61) explicit:

$$\widetilde{\mathbb{E}}\left[\left(\frac{\mathrm{d}\mathbb{P}}{\mathrm{d}\widetilde{\mathbb{P}}}(\omega)\right)^{\frac{1}{1-\gamma}}\right]^{1-\gamma} = \left(\frac{\det(I-H)}{\left(\det(I-(1-\gamma)^{-1}H)\right)^{1-\gamma}}\right)^{1/2}, \tag{6.67}$$

and we point out that such a computation requires also $I-(1-\gamma)^{-1}H$ to be positive definite, which is a direct consequence of (6.66) which guarantees that the spectrum of $I-(1-\gamma)^{-1}H$ is bounded below by $1/2$. The right-hand side of (6.67) can be estimated by observing that

1. Since $1-x \le \exp(-x)$, $\det(I-H) = \prod_j (1-\lambda_j(H)) \le \exp(-\sum_j \lambda_j(H)) = 1$
2. Since $\log(1+x) \ge x - x^2$ for $x \ge -1/2$, by using (6.66) and, once again, that H is traceless we see that $\det(I-(1-\gamma)^{-1}H) \ge \exp(-(1-\gamma)^{-2}\|H\|^2) \ge 1/e$

so that

$$\widetilde{\mathbb{E}}\left[\left(\frac{\mathrm{d}\mathbb{P}}{\mathrm{d}\widetilde{\mathbb{P}}}(\omega)\right)^{\frac{1}{1-\gamma}}\right]^{1-\gamma} \le e^{1-\gamma} = e^{1/5} < 2. \tag{6.68}$$

This term is therefore under control: let us turn to the second factor in the right-hand side of (6.61). We start by observing

$$\begin{aligned}
\widetilde{\mathbb{E}}\left[Z_{k,\omega}\right] &= \mathbf{E}\widetilde{\mathbb{E}}\left[\exp\left(\sum_{n=1}^{k}\left(\beta\omega_n - \frac{\beta^2}{2}+h\right)\delta_n\right)\delta_k\right] \\
&= \mathbf{E}\left[\exp\left(-\beta^2 \sum_{1\le i<j\le k} H_{i,j}\delta_i\delta_j + h\sum_{i=1}^{k}\delta_i\right)\delta_k\right] \\
&\le e\mathbf{E}\left[\exp\left(-\beta^2 \sum_{1\le i<j\le k} H_{i,j}\delta_i\delta_j\right)\middle|\,\delta_k=1\right]\mathbf{P}(\delta_k=1),
\end{aligned} \tag{6.69}$$

where the inequality in the last step comes from bounding $\sum_{j=1}^{k}\delta_j$ with k and by using $kh \le 1$. In view of this we are done if we can show that for every $\eta > 0$ we can find $\mathsf{c} > 0$ such that for every $\beta \in (0, \beta_*]$ we have

$$\widetilde{\mathbb{E}}\left[Z_{k,\omega}\right] \le \frac{\eta^{1/\gamma}}{e}\mathbf{P}(\delta_k=1). \tag{6.70}$$

This of course follows if we can show that the expectation in the last line of (6.69) can be made small by choosing c large. This estimate is somewhat technical: let us start with a more straightforward observation that is not conclusive, but it goes in the right direction. Let us show that what is at the exponent in the expectation that we have to estimate is *large*: by using (A.11) one sees that

$$\begin{aligned}
\mathbf{E}\left[\beta^2 \sum_{1\le i<j\le k} H_{i,j}\delta_i\delta_j\,\middle|\,\delta_k=1\right] &= \beta^2 \sum_{1\le i<j\le k} H_{i,j}\mathbf{E}\left[\delta_i\delta_j\,|\,\delta_k=1\right] \\
\ge c\beta^2 \sum_{1\le i<j\le k}\frac{H_{i,j}}{\sqrt{i(j-i)}} &\ge \frac{c\beta^2(1-\gamma)}{3\sqrt{k}\log k}\sum_{i=1}^{k/2}\frac{1}{\sqrt{i}}\sum_{j=1}^{k/2}\frac{1}{j} \overset{k\to\infty}{\sim} \frac{c\beta^2(1-\gamma)\sqrt{2\log k}}{3},
\end{aligned} \tag{6.71}$$

where c is a positive constant that depends only on $K(\cdot)$. Overall we see that the term that we are estimating in (6.71) is bounded below by (a positive constant times) $\beta^2\sqrt{\log k}$ for k sufficiently large large (recall that k large for us means $\mathtt{c}$ small), that is by $1/\sqrt{\mathtt{c}}$. Therefore the expectation of the random variable in the exponent of the expectation we need to control is negative and large: this suggests the result but it does not allow to conclude. In order to conclude we need to show that the random variable itself is *always* large: this is in fact the case and we explain it by resorting to a crucial probability estimate that we prove in the appendix (Proposition A.7) on which we are going to comment here. If we set

$$X_n := \frac{1}{\sqrt{n}\log n} \sum_{1\le i<j\le n} \frac{\delta_i \delta_j}{\sqrt{j-i}}, \tag{6.72}$$

then Proposition A.7 implies that for every $\lambda > 0$

$$\lim_{n\to\infty} \mathbf{E}\left[\exp(-\lambda X_n)\right] = \mathbf{E}\left[\exp\left(-\lambda(2\pi)^{-3/2}\mathtt{c}_K^{-2}|Z|\right)\right], \tag{6.73}$$

where Z is a standard Gaussian random variable. Two key points have to be remarked in the last formula: the normalization $\log n$ (and not $\sqrt{\log n}$) in X_n, so that we obtain the case that interests us by choosing "$\lambda = \sqrt{\log n}$", and the fact that $|Z| > 0$ with probability one, so that the right-hand side can be made arbitrarily small for λ large.

Let us see how this result is applied: first of all we use Proposition A.5 to obtain

$$\mathbf{E}\left[\exp\left(-\beta^2 \sum_{1\le i<j\le k} H_{i,j}\delta_i\delta_j\right)\middle| k\in\tau\right] \le C_{bc}\,\mathbf{E}\left[\exp\left(-\beta^2 \sum_{1\le i<j\le k/2} H_{i,j}\delta_i\delta_j\right)\right], \tag{6.74}$$

and in turn

$$\mathbf{E}\left[\exp\left(-\beta^2 \sum_{1\le i<j\le k/2} H_{i,j}\delta_i\delta_j\right)\right] = \mathbf{E}\left[\exp\left(-\frac{1-\gamma}{3}\left(\beta^2\sqrt{\log k}\right)X_{k/2}\right)\right]. \tag{6.75}$$

Going back to (6.69) and to (6.70) we see that we want to make the quantity in (6.75) smaller that $\eta/(eC_{bc})$. Therefore choose $\lambda = \lambda_0$ such that the right-hand side of (6.73) is equal to $\eta/(2eC_{bc})$. Then choose $\mathtt{c}$ sufficiently small so that $(1-\gamma)$ $\beta^2\sqrt{\log k}/3 \ge \lambda_0$: this is possible because $\beta^2\sqrt{\log k} \sim \sqrt{1/\mathtt{c}}$ for $\mathtt{c}$ small. With these choices we have

$$\mathbf{E}\left[\exp\left(-\beta^2 \sum_{1\le i<j\le k/2} H_{i,j}\delta_i\delta_j\right)\right] \le \mathbf{E}\left[\exp\left(-\lambda_0 X_{k/2}\right)\right] \le \frac{\eta^{1/\gamma}}{eC_{bc}}, \tag{6.76}$$

where in the last step we have chosen k sufficiently large (i.e. $\mathtt{c}$ small) and we have used (6.73). This concludes the proof of (6.70). □

6.6.2 Beyond the Correlation Length: The Proof of Theorem 6.1 ($\alpha = 1/2$)

We treat separately the case of Gaussian charges and the general case.

The case of Gaussian charges. Verifying the three hypotheses of Proposition 7.1 is just a revisitation of the proof of Proposition 6.8. In particular we keep the very same definitions of k and h and $\widetilde{\mathbb{P}}_k$ of Proposition 7.1 is the centered Gaussian measure with covariance $I - H$ given in (6.63), that is $\widetilde{\mathbb{P}}_k = \widetilde{\mathbb{P}}$. Therefore Hypothesis (1) of Proposition 7.1 is just (6.68).

For Hypotheses (2) and (3) we compute [like in (6.69)] for $d, f \in \{1, \dots, k\}$ with $d \le f$

$$\widetilde{\mathbb{E}}_k \mathbf{E}\left[\exp\left(\sum_{n=d}^{f} (\beta\omega_n - \log \mathrm{M}(\beta) + h)\,\delta_n\right)\delta_f \middle| d \in \tau\right] =$$

$$\mathbf{E}\left[\exp\left(-\beta^2 \sum_{\substack{i,j \in \{d,\dots,f\}\\ i<j}} H_{i,j}\delta_i\delta_j + h\sum_{i=d}^{f}\delta_i\right)\middle| d, f \in \tau\right] \mathbf{P}(f - d \in \tau)\,. \qquad (6.77)$$

Hypothesis (3) therefore holds, say for $\varepsilon \le 1/2$, because in this case $h\sum_{i=d}^{f}\delta_i \le \varepsilon hk \le \varepsilon$ and $H_{i,j} > 0$. For Hypothesis (2) we use, like in (6.69), $h\sum_{i=d}^{f}\delta_i \le hk \le 1$, paying thus a factor e and one is left with estimating (note that $H_{i,j}$ is just a function of $i - j$)

$$\mathbf{E}\left[\exp\left(-\beta^2 \sum_{1 \le i < j \le f-d+1} H_{i,j}\delta_i\delta_j\right)\middle| f - d + 1 \in \tau\right]. \qquad (6.78)$$

If we now repeat steps (6.74) and (6.75) we see that the expression in (6.78) is bounded by C_{bc} times

$$\begin{aligned}
&\mathbf{E}\left[\exp\left(-\beta^2 \sum_{1 \le i < j \le (f-d+1)/2} H_{i,j}\delta_i\delta_j\right)\right] \\
&\qquad \le \mathbf{E}\left[\exp\left(-\frac{1-\gamma}{3}\beta^2\sqrt{\frac{\varepsilon}{2}}\left(\frac{\log(\varepsilon k/2)}{\sqrt{\log k}}\right) X_{\lfloor \varepsilon k/2 \rfloor}\right)\right] \\
&\qquad\qquad \le \mathbf{E}\left[\exp\left(-\frac{\sqrt{\varepsilon}}{30}\left(\beta^2\sqrt{\log k}\right) X_{\lfloor \varepsilon k/2 \rfloor}\right)\right] \qquad (6.79)
\end{aligned}$$

for c sufficiently small. The last term we have obtained differs from the right-hand term in (6.75) only for the presence of ε: the argument therefore runs precisely like for (6.75) and this time c will have to be chosen suitably small in dependence of $K(\cdot)$, β_*, η and ε. □

The case of general charges. The generalization in this case is not straightforward because only in the Gaussian case one has directly the hands on correlations. The approach we choose here is the one in [20] and [25]. We point out that we do not aim at showing the full result proven in in [20], see Sect. 6.7. In any case, as for the $\alpha = 1/2$ Gaussian we set $\gamma = 4/5$ and $k(\beta, \mathrm{c})$ si still given by (6.59), with $h = \exp(-1/(\mathrm{c}\beta^4))$.

We introduce the random variable

$$Y = Y(\omega; k) := \sum_{0<i<j\le k} V_k(j-i)\omega_i\omega_j, \tag{6.80}$$

with $V_k(n) = (nk\log k)^{-1/2}$. Note that

$$\mathbb{E}Y = 0 \quad \text{and} \quad \mathbb{E}\left[Y^2\right] = \sum_{0<i<j\le k} (V_k(j-i))^2 \stackrel{k\to\infty}{\sim} 1, \tag{6.81}$$

so that we can assume that the variance of Y is bounded by (say) two. Note also the closeness with (6.63). We now introduce

$$g_k(\omega) := \exp\left(-\gamma\mathrm{K}\mathbf{1}_{Y(\omega;k)\ge\exp(\mathrm{K}^2)}\right), \tag{6.82}$$

where $\mathrm{K} > 0$ is going to be chosen below. We now re-edit the step (6.61) as

$$\mathbb{E}\left[Z_{k,\omega}^\gamma\right] \le \mathbb{E}\left[g_k^{-1/(1-\gamma)}(\omega)\right]^{1-\gamma} \mathbb{E}\left[g_k^{1/\gamma} Z_{k,\omega}\right]^\gamma. \tag{6.83}$$

One of the differences with (6.61) is that $g_k(\cdot)$ is not a probability density, but in reality we could make it a probability density with no effort: in fact, because of (6.81), $\mathbb{P}(Y(\omega;k) \ge \exp(\mathrm{K}^2))$ can be made arbitrarily small by choosing K large, so that

$$\mathbb{E}\left[g_k(\omega)\right] = 1 - (1-\exp(-\gamma\mathrm{K}))\mathbb{P}(Y(\omega;k) \ge \exp(\mathrm{K}^2)), \tag{6.84}$$

which can be made larger than $1/2$ for K sufficiently large, so that we can normalize $g_k(\cdot)$ by paying an irrelevant multiplicative constant.

The first of the two factors on the right-hand side of (6.83) is quickly controlled:

$$\begin{aligned} \mathbb{E}\left[g_k^{-1/(1-\gamma)}(\omega)\right] &= 1 + (\exp(\gamma\mathrm{K}/(1-\gamma)) - 1)\mathbb{P}(Y(\omega;k) \ge \exp(\mathrm{K}^2)) \\ &\le 1 + (\exp(\gamma\mathrm{K}/(1-\gamma)) - 1)\exp(-2\mathrm{K}^2)\mathbb{E}\left[Y^2(\omega;k)\right], \end{aligned} \tag{6.85}$$

and the last term tends to 1 for $\mathrm{K} \to \infty$ by (6.81).

We therefore turn to controlling the second factor on the right-hand side of (6.83) and for this we introduce, given a realization of the renewal τ, the probability $\widehat{\mathbb{P}}_\tau$ such that

$$\frac{\mathrm{d}\widehat{\mathbb{P}}_\tau}{\mathrm{d}\mathbb{P}}(\omega) = \exp\left(\sum_{n=1}^{k} (\beta\omega_n - \log \mathrm{M}(\beta))\,\delta_n\right). \tag{6.86}$$

We recall that by our choice of h ($kh \le 1$), exactly like in (6.69), we have $Z_{k,\omega,\beta,h_c^a+h} \le eZ_{k,\omega,\beta,h_c^a}$ and it suffices to consider the $h=0$ case ($h=0$ in the partition function, not in the definition of k), that is to show that for every $\zeta > 0$ we have that for $h=0$

$$\frac{\mathbb{E}\left[g_k^{1/\gamma} Z_{k,\omega}\right]}{\mathbf{P}(k\in\tau)} \le \zeta, \tag{6.87}$$

by choosing $\mathtt{c}$ small, and this uniformly in $\beta \in (0,\beta_*]$, for some β_*. The technical instrument at this point is the following lemma.

Lemma 6.9. *We have that*

1. *There exists* $\mathtt{c}_0 > 0$ *such that for* $\mathtt{c} \le \mathtt{c}_0$ *we have*

$$\mathbf{E}\left[\widehat{\mathbb{E}}_\tau\left[\left(Y(\omega;k) - \widehat{\mathbb{E}}_\tau[Y(\omega;k)]\right)^2\right]\middle| k\in\tau\right] \le \frac{1}{\mathtt{c}^{1/2}}. \tag{6.88}$$

2. *For every* $\zeta_2 > 0$ *there exist* $a > 0$ *and* $\mathtt{c}_0 > 0$ *such that for* $\mathtt{c} \le \mathtt{c}_0$ *we have*

$$\mathbb{P}\left(\widehat{\mathbb{E}}_\tau[Y(\omega;k)] > a\mathtt{c}^{-1/2}\middle| k\in\tau\right) \ge 1-\zeta_2. \tag{6.89}$$

We postpone the proof and meanwhile we observe that

$$\begin{aligned}\frac{\mathbb{E}\left[g_k^{1/\gamma} Z_{k,\omega}\right]}{\mathbf{P}(k\in\tau)} &= \mathbf{E}\left[\exp\left(-\kappa \mathbf{1}_{Y\ge \exp(\kappa^2)}\right)\middle| k\in\tau\right] \\ &\le \exp(-\kappa) + \mathbf{E}\left[\widehat{\mathbb{P}}_\tau\left(Y(\omega;k) < \exp(\kappa^2)\right)\middle| k\in\tau\right],\end{aligned} \tag{6.90}$$

and we make the choice of $\kappa = \kappa(a,\mathtt{c})$:

$$2\exp(\kappa^2) = a\mathtt{c}^{-1/2}, \tag{6.91}$$

which guarantees that κ can be made large by choosing $\mathtt{c}$ small and, cf. Lemma 6.9, that

$$\mathbb{P}\left(\widehat{\mathbb{E}}_\tau[Y(\omega;k)] \le 2\exp(\kappa^2)\middle| k\in\tau\right) \ge \zeta_2. \tag{6.92}$$

Let us both set the value of ζ_2 equal to $\zeta/3$ and let us choose $\mathtt{c}$ sufficiently small to guarantee that $\exp(-\kappa) \le \zeta/3$. We therefore restart from (6.90) and continue the chain of inequalities

$$
\begin{aligned}
\frac{\mathbb{E}\left[g_k^{1/\gamma} Z_{k,\omega}\right]}{\mathbf{P}(k\in\tau)} &\le \frac{2}{3}\zeta + \mathbf{E}\left[\widehat{\mathbb{P}}_\tau\left(Y(\omega;k) - \widehat{\mathbb{E}}_\tau[Y(\omega;k)] < -\exp(\kappa^2)\right) \Big| k\in\tau\right] \\
&\le \frac{2}{3}\zeta + \frac{4\mathtt{c}}{a^2}\mathbf{E}\left[\widehat{\mathbb{E}}_\tau\left[\left(Y(\omega;k) - \widehat{\mathbb{E}}_\tau[Y(\omega;k)]\right)^2\right] \Big| k\in\tau\right] \\
&\le \frac{2}{3}\zeta + \frac{4\mathtt{c}^{1/2}}{a^2} \le \zeta,
\end{aligned}
\tag{6.93}
$$

where in the last steps we have applied Lemma 6.9(1) and we have chosen $\mathtt{c} \le \zeta^2 a^4/144$. This completes the argument, but of course the proof of Lemma 6.9 is still due.

Proof of Lemma 6.9. First of all we remark that the charges are still independent under the measure $\widehat{\mathbb{P}}_\tau$, but they are no longer identically distributed. In fact if we set $m_\beta := \mathbf{E}[\omega_1 \exp(\beta\omega_1)]/\mathbf{E}[\exp(\beta\omega_1)]$ we have $m_\beta \sim \beta$ when $\beta \searrow 0$ and

$$
\widehat{\mathbb{E}}_\tau[\omega_n] = m_\beta \delta_n\,,
\tag{6.94}
$$

along with the fact that $\widehat{\mathbb{E}}_\tau[\omega_n^2]$ is bounded for every $\beta \le \beta_*$ by a constant that depends only on β_* and the law of ω_1 (we call it $c(\beta_*)$).

Let us look then at part (1). We set $\widehat{\omega}_n := \omega_n - \delta_n m_\beta$ and compute

$$
\begin{aligned}
&\widehat{\mathbb{E}}_\tau\left[\left(Y(\omega;k) - \widehat{\mathbb{E}}_\tau[Y(\omega;k)]\right)^2\right] \\
&\quad = \widehat{\mathbb{E}}_\tau\left[\left(\sum_{i<j\le k} V_k(j-i)\left(\widehat{\omega}_i\widehat{\omega}_j + m_\beta\delta_i\widehat{\omega}_j + m_\beta\delta_j\widehat{\omega}_i\right)\right)^2\right] \\
&\le 3c(\beta_*)^2 \sum_{i<j\le k}(V_k(j-i))^2 + 3m_\beta^2 c(\beta_*)\sum_j\left(\sum_{i,i'<j} + \sum_{i,i'>j}\right) V_k(|j-i|)V_k(|j-i'|)\delta_i\delta_{i'},
\end{aligned}
\tag{6.95}
$$

where in the last step we have used the Cauchy–Schwarz inequality, independence and the bound on the variance of $\widehat{\omega}_n$. The first term of the last line is bounded by a constant uniformly in τ, cf. (6.81), while for the second we take the expectation with respect to $\mathbf{P}(\cdot|k\in\tau)$, use (A.11) and we perform the summation.

For what concerns part (2) we observe that

$$\widehat{\mathbb{E}}_\tau[Y(\omega;k)] = \frac{m_\beta^2}{\sqrt{k\log k}} \sum_{i<j\le k} \frac{\delta_i \delta_j}{\sqrt{j-i}}. \tag{6.96}$$

But if one recalls that $m_\beta \sim \beta$ and that $V_k(j-i)$ coincides (up to innocuous multiplicative constants) with $H_{i,j}$ one realizes that the estimate we are asking is precisely what we have developed in the case of Gaussian charges [cf. (6.71)–(6.76)]. □

6.7 A Look at the Literature

Before going into the *history* of Theorem 6.1 let us point out that it is *sharp* (at least for $\alpha \neq 1$: both these notes and the literature are quite lazy about the technically heavy and not so rewarding case $\alpha = 1$), in the sense that upper bounds that, up to multiplicative constants, match the lower bounds in (6.1) are proved in [1, 30].

The argument of proofs in this chapter are all revolving around fractional moment estimates on the partition function. Fractional moments bounds have been exploited in statistical mechanics, notably in the analysis of random Schrödinger operators [3], but their relevance for disordered pinning models has been pointed out by Toninelli in [31] and his argument is explained in Sect. 6.2. Let us point out that the method of *constrained annealing* [27] that is often very useful in improving on annealed bounds is of no use to show that $h_c(\beta) - h_c^a(\beta) > 0$ [12]. The proof of Theorem 6.1 is based on substantial refinements of the basic fractional moment approach. A result close to Theorem 6.1, at least for $\alpha > 1/2$, has been proven in [14], using precisely the line that we followed here for the case $\alpha \geq 1$. The results in [14] have been improved in [2], without using fractional moments, but still only for $\alpha > 1/2$ [in the sense of (2.30)]. However, in order to fully appreciate the results in [2, 14] one should take into account that they have been obtained for regularly varying functions $K(\cdot)$, in particular allowing $K(n) \sim (\log(n))^\zeta / n^{1+\alpha}$. So, both [14] and [2] deal with the case $\alpha = 1/2$, but they demand $\zeta < 0$ ([14] even $\zeta < -1/2$).

The case $\alpha = 1/2$ has been solved in [19] and then the result has been improved in [20]. The proof we present here for the case $\alpha \in [1/2, 1)$ is a simplification of the technique in [19, 20]. It is a simplification not only because there is no slowly varying function in $K(\cdot)$, that is $\zeta = 0$, but also because in [20] it is proven a result that is sensibly stronger than (6.1), $\alpha = 1/2$, namely that $\exp(-1/(c\beta^4))$ can be replaced by $\exp(-1/(c\beta^{2+a}))$, with $a > 0$. This

1. Arrives close to what is claimed in [15], where the critical point shift is quantified, to leading order, to $C_1 \exp(-C_2/(c\beta^2))$, with C_1 and C_2 explicit positive constants.
2. Requires upgrading the 2-body potential argument we have presented to q-body potentials, $q > 2$: the *2-body* refers to the fact that $Y(\omega;k)$ [cf. (6.80)] is a second degree polynomial.

Moreover in [20] it is shown that $h_c(\beta) > h_c^a(\beta)$, with an explicit bound, for $\alpha = 1/2$ and $\zeta < 1/2$. This is close to being the optimal result that one expects: disorder irrelevance has been in fact established in [1,30] (see also Chap. 4) as soon as $\sum_n \mathbf{P}(n \in \tau)^2 < \infty$, that is, for $\alpha = 1/2$, as soon as $\sum_n 1/(n(\log n)^{2\zeta}) < \infty$. The *only* case left out is $\zeta = 1/2$ (and, of course, all the other slowly varying functions thereabout...)!

The work [19] has actually settled the controversy on whether disorder is relevant or irrelevant at marginality, at least from the view point of critical point shift. After the appearance of [16, 21, 22] stood in favor of disorder irrelevance and $h_c(\beta) = h_c^a(\beta)$ for $\alpha = 1/2$ [in the sense of (2.30)] and β not too large, but once [15] was published with the claim $h_c(\beta) > h_c^a(\beta)$, the physical literature seemed to settle on the new line [6, 28, 29]. More recently however the claim [16] that $h_c(\beta) = h_c^a(\beta)$ has reappeared in [17].

The method adopted in this chapter (fractional moments, measure change, coarse graining) has been successfully used in other contexts. Here we mention

1. Copolymers at selective interfaces [11,32]
2. The pinning on a walk model [5, 8, 9]
3. The directed polymer in random environment [25, 26], in particular with a proof [25] of the fact that quenched and annealed free energies differ as soon as $\beta > 0$ in dimension one and, notably, two
4. Random walks in random environments [33]
5. Semi-directed polymers in random environment [34]

Recently another approach, based on the quenched large deviation principle proven in [10], to decide whether or not $h_c(\beta) > h_c^a(\beta)$ has been set forth in [13]. While for now this approach does not yield $h_c(\beta) > h_c^a(\beta)$ for $\alpha > 1/2$, it provides such a result for β sufficiently large, and arbitrary α, under a condition on the disorder that is weaker than what one can extract form Proposition 6.3. The condition is easily stated if we call w the supremum of the support of the law of ω_1: it is simply

$$\mathbb{P}(\omega_1 = w) = 0, \tag{6.97}$$

and this includes all the cases in which $w = \infty$. For $w < \infty$ this fact is better appreciated in relation to Theorem 4.5 and (4.11).

We cannot close this chapter without recalling that much of the developments and ideas presented in this chapter were born while dealing with the hierarchical version of disordered pinning (a model proposed in [15]). Hierarchical models did not find their place in these notes, but we do want to mention [18, 19, 24].

References

1. K.S. Alexander, The effect of disorder on polymer depinning transitions. Commun. Math. Phys. **279**, 117–146 (2008)
2. K.S. Alexander, N. Zygouras, Quenched and annealed critical points in polymer pinning models. Commun. Math. Phys. **291**, 659–689 (2009)

3. M. Aizenman, S. Molchanov, Localization at large disorder and at extreme energies: an elementary derivation. Commun. Math. Phys. **157**, 245–278 (1993)
4. Q. Berger, H. Lacoin, The effect of disorder on the free-energy for the random walk pinning model: smoothing of the phase transition and low temperature asymptotics. J. Stat. Phys. **42**, 322–341 (2011)
5. Q. Berger, F.L. Toninelli, On the critical point of the random walk pinning model in dimension $d = 3$. Electron. J. Probab. **15**, 654–683 (2010)
6. S.M. Bhattacharjee, S. Mukherji, Directed polymers with random interaction: marginal relevance and novel criticality. Phys. Rev. Lett. **70**, 49–52 (1993)
7. N.H. Bingham, C.M. Goldie, J.L. Teugels, *Regular Variation* (Cambridge University Press, Cambridge, 1987)
8. M. Birkner, R. Sun, Annealed vs quenched critical points for a random walk pinning model. Ann. Inst. H. Poincaré (B) Probab. Stat. **46**, 414–441 (2010)
9. M. Birkner, R. Sun, Disorder relevance for the random walk pinning model in dimension 3. Ann. Inst. H. Poincaré (B) Probab. Stat. **47**, 259–293 (2011)
10. M. Birkner, A. Greven, F. den Hollander, Quenched large deviation principle for words in a letter sequence. Probab. Theory Relat. Fields **148**, 403–456 (2010)
11. T. Bodineau, G. Giacomin, H. Lacoin, F.L. Toninelli, Copolymers at selective interfaces: new bounds on the phase diagram. J. Stat. Phys. **132**, 603–626 (2008)
12. F. Caravenna, G. Giacomin, On constrained annealed bounds for pinning and wetting models. Electron. Commun. Probab. **10**, 179–189 (2005)
13. D. Cheliotis, F. den Hollander, Variational characterization of the critical curve for pinning of random polymers. arXiv:1005.3661
14. B. Derrida, G. Giacomin, H. Lacoin, F.L. Toninelli, Fractional moment bounds and disorder relevance for pinning models. Commun. Math. Phys. **287**, 867–887 (2009)
15. B. Derrida, V. Hakim, J. Vannimenus, Effect of disorder on two-dimensional wetting. J. Stat. Phys. **66**, 1189–1213 (1992)
16. G. Forgacs, J.M. Luck, Th. M. Nieuwenhuizen, H. Orland, Wetting of a disordered substrate: exact critical behavior in two dimensions. Phys. Rev. Lett. **57**, 2184–2187 (1986)
17. D.M. Gangardt, S.K. Nechaev, Wetting transition on a one-dimensional disorder. J. Stat. Phys. **130**, 483–502 (2008)
18. G. Giacomin, H. Lacoin, F.L. Toninelli, Hierarchical pinning models, quadratic maps and quenched disorder. Probab. Theory Relat. Fields **147**, 185–216 (2010)
19. G. Giacomin, H. Lacoin, F.L. Toninelli, Marginal relevance of disorder for pinning models. Commun. Pure Appl. Math. **63**, 233–265 (2010)
20. G. Giacomin, H. Lacoin, F.L. Toninelli, Disorder relevance at marginality and critical point shift. Ann. Inst. H. Poincaré (B) Probab. Stat. **47**, 148–175 (2011)
21. A.Y. Grosberg, E.I. Shakhnovich, An investigation of the configurational statistics of a polymer chain in an external field by the dynamical renormalization group method. Sov. Phys. JETP **64**, 493–501 (1986)
22. A.Y. Grosberg, E.I. Shakhnovich, Theory of phase transitions of the coil-globule type in a heteropolymer chain with disordered sequence of links. Sov. Phys. JETP **64**, 1284–1290 (1986)
23. A.B. Harris, Effect of random defects on the critical behaviour of Ising models. J. Phys. C **7**, 1671–1692 (1974)
24. H. Lacoin, Hierarchical pinning model with site disorder: disorder is marginally relevant. Probab. Theory Relat. Fields **148**, 159–175 (2010)
25. H. Lacoin, New bounds for the free energy of directed polymer in dimension 1+1 and 1+2. Commun. Math. Phys. **294**, 471–503 (2010)
26. H. Lacoin, Influence of spatial correlation for directed polymers. Ann. Probab. **39**, 139–175 (2011)
27. T. Morita, Statistical mechanics of quenched solid solutions with application to magnetically dilute alloys. J. Math. Phys. **5**, 1401–1405 (1966)
28. S. Stepanow, A.L. Chudnovskiy, The Green's function approach to adsorption of a random heteropolymer onto surfaces. J. Phys. A Math. Gen. **35**, 4229–4238 (2002)

29. L.-H. Tang, H. Chaté, Rare-event induced binding transition of heteropolymers. Phys. Rev. Lett. **86**, 830–833 (2001)
30. F.L. Toninelli, A replica-coupling approach to disordered pinning models. Commun. Math. Phys. **280**, 389–401 (2008)
31. F.L. Toninelli, Disordered pinning models and copolymers: beyond annealed bounds. Ann. Appl. Probab. **18**, 1569–1587 (2008)
32. F.L. Toninelli, Coarse graining, fractional moments and the critical slope of random copolymers. Electron. J. Probab. **14**, 531–547 (2009)
33. A. Yilmaz, O. Zeitouni, Differing averaged and quenched large deviations for random walks in random environments in dimensions two and three. Commun. Math. Phys. (to appear). arXiv:0910.1169
34. N. Zygouras, Strong disorder in semidirected random polymers. arXiv:1009.2693

Chapter 7
The Coarse Graining Procedure

Abstract This chapter develops in detail the most advanced of the two coarse graining techniques employed in the previous chapter. Roughly, it consists in looking at the system in blocks of finite size k, which essentially is the annealed correlation length: if we can get suitable estimates for systems up to that size k, we can bound the fractional moment of the partition function of the (arbitrarily large) system in terms of the partition function of a homogeneous model with pinning parameter that depend on the estimates up to size k.

7.1 Coarse Graining Estimates

This is a technical chapter that details how suitable upper bounds on the finite volume system can be upgraded to an upper bound on arbitrarily large volumes. As the title says, it details a suitable coarse graining technique (it is adapted from [2], that proposes an improved version of the coarse graining in [1], that in turns was adapted from the coarse graining in [3] that deals with the copolymer model). We will use (only in this chapter) the compact notation

$$\omega_{\beta,n} := \beta\beta_n - \log \mathrm{M}(\beta), \tag{7.1}$$

and for $0 \le M < N$ we set

$$Z_{M,N} = Z_{M,N,\omega} := \mathbf{E}\left[\exp\left(\sum_{n=M+1}^{N} (\omega_{\beta,n} + h)\, \delta_n \right) \delta_N \,\middle|\, \delta_M = 1 \right], \tag{7.2}$$

and $Z_{M,M} := 1$. Moreover $Z_N := Z_{0,N}$.

In the statement that follows $k = k(\beta, \mathtt{c}) \in \mathbb{N}$ is a correlation length that is function of β ($\beta \in (0, \beta_*]$, where β_* is an arbitrary value that we keep fixed) and of a positive constant $\mathtt{c}$. What we are assuming is that

$$\lim_{\mathtt{c}\searrow 0} \inf_{\beta\in(0,\beta_*]} k(\beta, \mathtt{c}) = \infty. \tag{7.3}$$

G. Giacomin, *Disorder and Critical Phenomena Through Basic Probability Models*, Lecture Notes in Mathematics 2025, DOI 10.1007/978-3-642-21156-0_7,

© Springer-Verlag Berlin Heidelberg 2011

As a matter of fact, in all the cases we consider the correlation length is completely explicit, it is either (the integer part of) $\mathtt{c}^{-c_1}\beta_*^{-c_2}$ with $c_1, c_2 > 0$ or $\exp(1/\mathtt{c}\beta_*^4)$, and $\inf_{\beta\in(0,\beta_*]} k(\beta,\mathtt{c}) = k(\beta_*,\mathtt{c})$. We also need to choose a dependence of h on $\mathtt{c}$ and β. For the intuition it is more convenient to think and write h as a (positive) function the correlation length: $h = h(k) > 0$, and in the applications $h(k)$ is just k to some negative power. To be precise, in the applications we will not keep the integer part restrictions on k when dealing with h, so we will use rather $h(\mathtt{c},\beta)$ rather than $h(k)$, but this is really inessential.

The value of $\alpha > 0$ is arbitrary, but the cases in which we are interested in is $\alpha \in [1/2, 1)$: we assume $\alpha \in (0,1)$ because on one hands some constants do depend on α so that one has to work with α smaller than a constant and because the renewal function estimates change at $\alpha = 1$ and working in the general set-up would be unnecessarily heavy. Moreover we set

$$\gamma := \frac{2}{2+\alpha} \in (0,1), \tag{7.4}$$

so that $(1+\alpha)\gamma > 1$.

Proposition 7.1. *Let us choose $k(\cdot,\cdot)$, $h(\cdot)$, α and γ as above and let us assume that for every $\eta, \varepsilon \in (0,1)$ one can find $\mathtt{c}_0 > 0$ and $\left\{\widetilde{\mathbb{P}}_n\right\}_{n\in\mathbb{N}}$, with $\widetilde{\mathbb{P}}_n$ a probability on $\mathbb{R}^n$ equivalent (i.e. mutually absolutely continuous) to the restriction $\mathbb{P}_n$ of $\mathbb{P}$ to $\mathbb{R}^n$, such that for every $\mathtt{c} \le \mathtt{c}_0$ and every $\beta \in (0,\beta_*]$ we have*

1.

$$\widetilde{\mathbb{E}}_k\left[\left(\frac{\mathrm{d}\mathbb{P}_k}{\mathrm{d}\widetilde{\mathbb{P}}_k}\right)^{1/(1-\gamma)}\right] \le 2, \tag{7.5}$$

2.

$$\sup_{\substack{f,d\in\{1,2,\ldots k\}\\ f-d>\varepsilon k}} \frac{\widetilde{\mathbb{E}}_k \mathbf{E}\left[\exp\left(\sum_{n=d}^{f}\left(\omega_{\beta,n}+h(k)\right)\delta_n\right)\delta_f \middle| d\in\tau\right]}{\mathbf{P}(f-d\in\tau)} \le \eta, \tag{7.6}$$

3.

$$\sup_{\substack{f,d\in\{1,2,\ldots k\}\\ f-d\le\varepsilon k}} \frac{\widetilde{\mathbb{E}}_k \mathbf{E}\left[\exp\left(\sum_{n=d}^{f}\left(\omega_{\beta,n}+h(k)\right)\delta_n\right)\delta_f \middle| d\in\tau\right]}{\mathbf{P}(f-d\in\tau)} \le 2. \tag{7.7}$$

Then there exists $\mathtt{c}_1 > 0$ such that for every $\mathtt{c} \in (0,\mathtt{c}_1]$, every $\beta \in (0,\beta_*]$ and for $h \le h(k)$ we have

$$\sup_N \mathbb{E}\left[Z_N^\gamma\right] < \infty. \tag{7.8}$$

Proof. Without loss of generality k is chosen even and $N = km$ ($m \in \mathbb{N}$ is therefore the size of the coarse grained system). A coarse grained configuration is a subset $\mathscr{I}$ of $\{1,2,\ldots,m\}$ and the coarse graining blocks, or k-blocks, are

$$B_i := (i-1)k + \{1,2,\ldots,k\}, \tag{7.9}$$

for $i = 1, 2, \ldots$. We introduce

$$\widehat{Z}_{\mathscr{I}} = \widehat{Z}_{\mathscr{I},\omega} := \mathbf{E}\left[\exp\left(\sum_{n=1}^{N}(\omega_{\beta,n}+h)\,\delta_n\right)\mathbf{1}_{\tau\cap(0,N]\subset\cup_{i\in\mathscr{I}}B_i,\,\tau\cap B_i\neq\emptyset \text{ for } i\in\mathscr{I}}\,\delta_N\right], \tag{7.10}$$

so that

$$Z_N = \sum_{\mathscr{I}} \widehat{Z}_{\mathscr{I}}, \tag{7.11}$$

where $\sum_{\mathscr{I}}$ stands for the summation over $\mathscr{I} \subset \{1, \ldots, m\}$ and we can as well restrict the summation to the $\mathscr{I}$'s that contain m, since $\widehat{Z}_{\mathscr{I}} = 0$ otherwise. By writing $\mathscr{I} = \{i_1, \ldots, i_\ell\}$, with $1 \le i_1 < i_2 < \ldots < i_\ell = m$, and by setting $z_n = \exp(\omega_{\beta,n} + h)$ we have the formula

$$\widehat{Z}_{\mathscr{I}} = \sum_{\substack{d_1,f_1\in B_{i_1}\\ d_1\le f_1}} \sum_{\substack{d_2,f_2\in B_{i_2}\\ d_2\le f_2}} \cdots \sum_{\substack{d_{\ell-1},f_{\ell-1}\in B_{i_{\ell-1}}\\ d_{\ell-1}\le f_{\ell-1}}} \sum_{d_\ell\in B_\ell}$$

$$K(d_1)z_{d_1}Z_{d_1,f_1}K(d_2-f_1)z_{d_2}Z_{d_2,f_2}\cdots z_{d_{\ell-1}}Z_{d_{\ell-1},f_{\ell-1}}K(d_\ell-f_{\ell-1})z_{d_\ell}Z_{d_\ell,N}. \tag{7.12}$$

For $\gamma \in (0,1)$ we have

$$\mathbb{E}\left[Z_N^\gamma\right] \le \sum_{\mathscr{I}} \mathbb{E}\left[\widehat{Z}_{\mathscr{I}}^\gamma\right]. \tag{7.13}$$

Given a coarse grained configuration $\mathscr{I}$ we now consider the new measure of the environment $\widetilde{\mathbb{P}}_{\mathscr{I}}$ with the following properties:

(P1) The law of $\{\omega_n\}_{n\notin\cup_{i\in\mathscr{I}}B_i}$ is the same under $\mathbb{P}$ or under $\widetilde{\mathbb{P}}_{\mathscr{I}}$, in particular they are still IID random variables.
(P2) The random variables $\{\omega_n\}_{n\in B_i}$ are independent of $\{\omega_n\}_{n\in B_{i'}}$ for $i \neq i'$.
(P3) The law of $\{\omega_j\}_{j\in B_i}$ coincides with the law of $\{\omega_j\}_{j\in B_{i'}}$ if $i, i' \in \mathscr{I}$.
(P3) The law of $\{\omega_j\}_{j\in B_i}$ for $i \in \mathscr{I}$, a probability on $\mathbb{R}^k$, is given by $\widetilde{\mathbb{P}}_k$ [the probability measure in the assumptions (7.5) and (7.6)].

The new environment is therefore changed only in the k-blocks that are *visited* in the coarse grained configuration (that is B_i such that $i \in \mathscr{I}$). We now apply the Hölder inequality with $p = 1/\gamma$ to get

$$\mathbb{E}\left[\widehat{Z}_{\mathscr{I}}^\gamma\right] \le \widetilde{\mathbb{E}}_{\mathscr{I}}\left[\left(\frac{\mathrm{d}\mathbb{P}}{\mathrm{d}\widetilde{\mathbb{P}}_{\mathscr{I}}}\right)^{1/(1-\gamma)}\right]^{1-\gamma} \mathbb{E}\left[\widehat{Z}_{\mathscr{I}}\right]^\gamma. \tag{7.14}$$

By the properties (P1) to (P4) above and by (7.5) we directly see that

$$\widetilde{\mathbb{E}}_{\mathscr{I}}\left[\left(\frac{\mathrm{d}\mathbb{P}}{\mathrm{d}\widetilde{\mathbb{P}}_{\mathscr{I}}}\right)^{1/(1-\gamma)}\right] = \widetilde{\mathbb{E}}_k\left[\left(\frac{\mathrm{d}\mathbb{P}_k}{\mathrm{d}\widetilde{\mathbb{P}}_k}\right)^{1/(1-\gamma)}\right]^{|\mathscr{I}|} \le 2^{|\mathscr{I}|}. \tag{7.15}$$

For the other term observe that by (7.12) and by property (P2) we have

$$\widetilde{\mathbb{E}}_{\mathscr{I}}\left[\widehat{Z}_{\mathscr{I}}\right] = \sum_{\substack{d_1,f_1\in B_{i_1}\\ d_1\le f_1}} \sum_{\substack{d_2,f_2\in B_{i_2}\\ d_2\le f_2}} \cdots \sum_{\substack{d_{\ell-1},f_{\ell-1}\in B_{i_{\ell-1}}\\ d_{\ell-1}\le f_{\ell-1}}} \sum_{d_\ell\in B_\ell}$$

$$K(d_1)\widetilde{\mathbb{E}}_{\mathscr{I}}\left[z_{d_1}Z_{d_1,f_1}\right] K(d_2-f_1)\widetilde{\mathbb{E}}_{\mathscr{I}}\left[z_{d_2}Z_{d_2,f_2}\right]\cdots K(d_\ell-f_{\ell-1})\widetilde{\mathbb{E}}_{\mathscr{I}}\left[z_{d_\ell}Z_{d_\ell,N}\right], \tag{7.16}$$

and properties (P3) and (P4), together with (7.6) and (7.7), tell us that

$$\widetilde{\mathbb{E}}_{\mathscr{I}}\left[z_{d_j}Z_{d_j,f_j}\right] = \widetilde{\mathbb{E}}_k\left[z_{d_j-(i_j-1)k}Z_{d_j-(i_j-1)k,f_j-(i_j-1)k}\right] \le \widetilde{G}(f_j-d_j), \tag{7.17}$$

with

$$\widetilde{G}(n) := (\eta + 2\mathbf{1}_{n\le\varepsilon k})\,\mathbf{P}(n\in\tau), \tag{7.18}$$

so that

$$\widetilde{\mathbb{E}}_{\mathscr{I}}\left[\widehat{Z}_{\mathscr{I}}\right] \le \sum_{\substack{d_1,f_1\in B_{i_1}\\ d_1\le f_1}} \sum_{\substack{d_2,f_2\in B_{i_2}\\ d_2\le f_2}} \cdots \sum_{\substack{d_{\ell-1},f_{\ell-1}\in B_{i_{\ell-1}}\\ d_{\ell-1}\le f_{\ell-1}}} \sum_{d_\ell\in B_\ell}$$

$$K(d_1)\widetilde{G}(f_1-d_1)K(d_2-f_1)\widetilde{G}(f_2-d_2)\cdots K(d_\ell-f_{\ell-1})\widetilde{G}(N-d_\ell). \tag{7.19}$$

We now aim at simplifying this expression by showing that if ε is chosen suitably (as a function of η) then we can replace $\widetilde{G}(n)$ in (7.19) with $2\eta\mathbf{P}(n\in\tau)$ and obtain an upper bound. It is rather intuitive why this is true: consider the jth visited block and think of the case in which d_j is not too close to the right-end of the block (say, it is in the first half of the block). Then in $\sum_{f=d_j}^{d_j+\varepsilon k}\mathbf{P}(f-d_j\in\tau)K(d_{j+1}-f)$ the term $K(d_{j+1}-f)$ depends little on f, since $d_{j+1}-f$ is larger than (say) $k/3$ (if ε is sufficiently small, say smaller than $1/10$). But $\sum_{f=d_j}^{d_j+\varepsilon k}\mathbf{P}(f-d_j\in\tau)$ for k large behaves like $(\varepsilon k)^\alpha$ [by (A.8)] while the analogous term if we remove the constraint of summing only over up to a distance εk we get $\sqrt{k}$: but this is what we are doing with the term in $\widetilde{G}(\cdot)$ that contains the factor η. So by choosing $\sqrt{\varepsilon}$ of the order of η we see that these two terms are comparable. Of course d_j is not always in the first half of the block, but if it is not, f_j is in the second half and a specular argument applies.

In detail, the argument goes as follows: let us consider one of the jth visited block and let us partition it into the left and the right part:

$$\begin{aligned} B^-_{i_j} &:= (i_j-1)k + \{1,\ldots,k/2\},\\ B^+_{i_j} &:= (i_j-1)k + \{k/2,\ldots,k\}. \end{aligned} \tag{7.20}$$

If $d_j \in B^-_{i_j}$ then for $\varepsilon < 1/12$ we have that for k sufficiently large ($k \geq k_0(K(\cdot), \varepsilon)$)

$$\sum_{f=d_j}^{d_j+\varepsilon k} \mathbf{P}(f - d_j \in \tau) K(d_{j+1} - f) \leq 6 \left(\sum_{n=1}^{k\varepsilon} \mathbf{P}(n \in \tau) \right) K\left((i_{j+1} - i_j) k \right). \quad (7.21)$$

This comes from the fact that when $i_{j+1} = i_j + 1$ then $d_{j+1} - f \geq (1/2 - 1/12)k$ and, since $K(5n/12) \overset{n\to\infty}{\sim} (12/5)^{1+\alpha} K(n)$ and $(5/12)^2 > 1/6$, we have that, for k large, $K(d_{j+1} - f) \leq 6K(k)$. If $i_{j+1} - i_j > 1$ the same argument can be repeated (with better constants). We compare this expression with the one in which we sum till the end of the block: for k sufficiently large ($k \geq k_0(K(\cdot))$) we have

$$\sum_{f=d_j}^{ki_j} \mathbf{P}(f - d_j \in \tau) K(d_{j+1} - f) \geq \frac{1}{4} \left(\sum_{n=1}^{k/4} \mathbf{P}(n \in \tau) \right) K\left((i_{j+1} - i_j) k \right). \quad (7.22)$$

Since $\sum_{n=1}^{N} \mathbf{P}(n \in \tau) \sim const. N^\alpha$ for N large we therefore see that by choosing for example $\varepsilon = \eta^3$ (of course η^2 times a small constant would suffice) we have that for η sufficiently small and for k sufficiently large ($k \geq k_0(K(\cdot), \eta)$)

$$\sum_{f=d_j}^{d_j+\varepsilon k} \mathbf{P}(f - d_j \in \tau) K(d_{j+1} - f) \leq \frac{\eta}{2} \sum_{f=d_j}^{ki_j} \mathbf{P}(f - d_j \in \tau) K(d_{j+1} - f). \quad (7.23)$$

If $d_j \notin B^-_{i_j}$ then $f_j \in B^+_{i_j}$ and one repeats the argument *on the left* to get

$$\sum_{d=f_j-\varepsilon k}^{f_j} K(d - f_{j-1}) \mathbf{P}(f_j - d \in \tau) \leq \frac{\eta}{2} \sum_{d=k(i_j-1)}^{f_j} K(d - f_{j-1}) \mathbf{P}(f_j - d \in \tau). \quad (7.24)$$

Therefore by (7.23) and (7.24) we see that

$$\sum_{\substack{d_j, f_j \in B_{i_j} \\ d_j \leq f_j}} \mathbf{1}_{\{f_j - d_j \leq k\varepsilon\}} K(d_j - f_{j-1}) \mathbf{P}(f_j - d_j \in \tau) K(d_{j+1} - f_j)$$

$$\leq \frac{\eta}{2} \sum_{\substack{d_j, f_j \in B_{i_k} \\ d_j \leq f_j}} K(d_j - f_{j-1}) \mathbf{P}(f_j - d_j \in \tau) K(d_{j+1} - f_j). \quad (7.25)$$

The last summation in (7.19) is slightly different, since one is summing only over d_ℓ. But in this case d_ℓ is necessarily very close to the right-end point of the block, and therefore far from f_{j-1} and the estimate requires only *half* of the argument that we have developed.

We sum up what we have obtained:

$$\widetilde{\mathbb{E}}_{\mathscr{I}}\left[\widehat{Z}_{\mathscr{I}}\right] \le (2\eta)^{\ell} \sum_{\substack{d_1,f_1\in B_{i_1}\\ d_1\le f_1}} \sum_{\substack{d_2,f_2\in B_{i_2}\\ d_2\le f_2}} \cdots \sum_{\substack{d_{\ell-1},f_{\ell-1}\in B_{i_{\ell-1}}\\ d_{\ell-1}\le f_{\ell-1}}} \sum_{d_\ell\in B_\ell}$$

$$K(d_1)\mathbf{P}(f_1-d_1\in\tau)K(d_2-f_1)\mathbf{P}(f_2-d_2\in\tau)\cdots K(d_\ell-f_{\ell-1})\mathbf{P}(N-d_\ell\in\tau)$$

$$= (2\eta)^{|\mathscr{I}|}\mathbf{P}(E(\mathscr{I})), \quad (7.26)$$

where $E(\mathscr{I})$ is the event that the coarse grained version of τ coincides with $\mathscr{I}$, namely:

$$E(\mathscr{I}) := \{\tau : \tau\cap(0,N] = \tau\cap(\cup_{i\in\mathscr{I}}B_i),\, \tau\cap B_i \neq \emptyset \text{ for } i\in\mathscr{I}\}. \quad (7.27)$$

We postpone the proof of following probability estimate.

Lemma 7.2. *There exist* $C_1 = C_1(K(\cdot),k)$, $C_2 = C_2(K(\cdot))$ *and* $k_0 = k_0(K(\cdot))$ *such that for* $k \ge k_0$

$$\mathbf{P}(E(\mathscr{I})) \le C_1 C_2^{|\mathscr{I}|} \prod_{j=1}^{|\mathscr{I}|} \frac{1}{(i_j - i_{j-1})^{1+\alpha}}, \quad (7.28)$$

where $i_0 := 0$.

Now we apply Lemma 7.2 to (7.26): in view of (7.13)–(7.15) we have

$$\mathbb{E}\left[Z_N^\gamma\right] \le C_1 \sum_{\mathscr{I}} 2^{(1-\gamma)|\mathscr{I}|}(2\eta C_2)^{\gamma|\mathscr{I}|} \prod_{j=1}^{|\mathscr{I}|} \frac{1}{(i_j-i_{j-1})^{(1+\alpha)\gamma}}$$

$$= C_1 \sum_{\mathscr{I}} (2^{1/\gamma}\eta C_2 A)^{\gamma|\mathscr{I}|} \prod_{j=1}^{|\mathscr{I}|} \widetilde{K}(i_j - i_{j-1}), \quad (7.29)$$

where $A := \sum_{n\in\mathbb{N}} n^{-(1+\alpha)\gamma}$ (recall that $(1+\alpha)\gamma > 1$) and $\widetilde{K}(n) := n^{-(1+\alpha)\gamma}/A$. With a different notation:

$$\mathbb{E}\left[Z_{mk}^\gamma\right] \le C_1\,\mathbf{E}\left[\exp(q\,|\widetilde{\tau}\cap(0,m]|);\, m\in\widetilde{\tau}\right], \quad (7.30)$$

where $\widetilde{\tau}$ is the renewal with inter-arrival law $\widetilde{K}(\cdot)$ and $q := \gamma\log(2^{1/\gamma}\eta C_2 A)$. By choosing $\eta < 2^{-1/\gamma}/(C_2 A)$ (that is η smaller than a quantity that depends only on $K(\cdot)$) we have in the right-hand side of (7.30) the partition function of a homogeneous pinning model in the delocalized regime. This tells us, in particular, that $\sup_m \mathbb{E}\left[Z_{mk}^\gamma\right] < \infty$ from which we directly [cf. (3.29)] infer $\sup_N \mathbb{E}\left[Z_N^\gamma\right] < \infty$ and we are done. □

Proof of Lemma 7.2. We start by observing that, by choosing $C_2 \geq 1$, we see that in the product on the right-hand side of (7.28) we can ignore the j's for which $i_j - i_{j-1} = 1$. We therefore prefer to reparametrize $\mathscr{I}$ in terms of two sequences of increasing integer numbers $\{a_j\}_{j=1,\dots,p}$, $\{b_j\}_{j=1,\dots,p}$ with $b_p = m$, $a_j \geq b_{j-1}+2$ (for $j > 1$) and $b_j \geq a_j$ such that

$$\mathscr{I} = \bigcup_{j=1}^{p} [a_j, b_j] \cap \mathbb{N}. \tag{7.31}$$

Of course $p \leq |\mathscr{I}|$.

With this definition, it is sufficient to show

$$\mathbf{P}(E(\mathscr{I})) \leq C_1 C_2^l \frac{1}{a_1^{1+\alpha}} \prod_{j=1}^{p-1} \frac{1}{(a_{j+1} - b_j)^{1+\alpha}}. \tag{7.32}$$

We start then by observing that $\mathbf{P}(E(\mathscr{I}))$ is bounded above by

$$\sum_{\substack{d_1 \in B_{a_1} \\ f_1 \in B_{b_1}}} \cdots \sum_{\substack{d_{p-1} \in B_{a_{p-1}} \\ f_p \in B_{b_{p-1}}}} \sum_{d_p \in B_{a_p}} K(d_1)\mathbf{P}(f_1 - d_1 \in \tau) \dots K(d_p - f_{p-1})\mathbf{P}(N - d_p \in \tau), \tag{7.33}$$

where the inequality comes from neglecting the fact that, when $b_j \geq a_j + 2$, a configuration $\tau \in E(\mathscr{I})$ is required to have a non-empty intersection with B_i for every $i \in \{a_j + 1, \dots, b_j - 1\}$. Note that the meaning of the d and f indexes is somewhat different with respect to (7.12) and that in the above sum we always have

$$\begin{aligned} (a_j - b_{j-1} - 1)k \leq d_j - f_{j-1} \leq (a_j - b_{j-1} + 1)k, \\ (b_j - a_j - 1)k \vee 0 \leq f_j - d_j \leq (b_j - a_j + 1)k. \end{aligned} \tag{7.34}$$

Notice also that, in (7.33), $f_j \geq d_j$ is granted, since $\mathbf{P}(f_j - d_j \in \tau) = 0$ if $f_j < d_j$.

By using $K(n) \leq C_0 n^{-1-\alpha}$ ($C_0 = C_0(K(\cdot)) > 0$) we directly get

$$\sum_{n \in B_{a_1}} K(n) \leq \begin{cases} 1 & \text{if } a_1 = 1, \\ C_0 k^{-\alpha}(a_1 - 1)^{-1-\alpha} & \text{if } a_1 = 2, 3, \dots, \end{cases} \leq \frac{c_1(k)}{k^\alpha a_1^{1+\alpha}}, \tag{7.35}$$

where $c_1(k) := \max(k^\alpha, C_0 2^{1+\alpha})$. Moreover for $j > 1$, since $a_j \geq b_{j-1} + 2$, we have

$$\begin{aligned} \sum_{n=(a_j - b_{j-1} - 1)k}^{(a_j - b_{j-1} + 1)k} K(n) &\leq \frac{2kC_0}{k^{1+\alpha}(a_j - b_{j-1} - 1)^{1+\alpha}} \\ &\leq \frac{2^{2-\alpha} C_0}{k^\alpha (a_j - b_{j-1})^{1+\alpha}} =: \frac{c_2}{k^\alpha (a_j - b_{j-1})^{1+\alpha}}, \end{aligned} \tag{7.36}$$

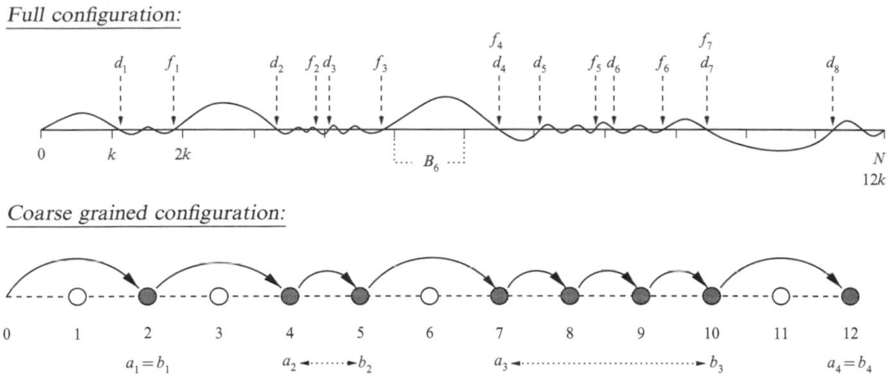

Fig. 7.1 *The coarse graining procedure*. At the top level a full configuration is given (a continuous path is drawn because it helps in identifying the contact points) and the k-blocks are marked. The indexes d_j and f_j of the decomposition (7.12) indicate respectively the first and the last renewal point in the jth visited block. The coarse grained configuration $\mathscr{I} \subset \{1,2,\dots,N/k=12\}$, below, is the one that keeps into account only the visited blocks. For the trajectory in the figure we have $\mathscr{I} = \{2,4,5,7,8,9,10,12\}$. In the last line one finds the parametrization of $\mathscr{I}$ used in Lemma 7.2, that consists in looking at the coarse grained configuration by *joining* the blocks, now big dots, that are neighbors

where c_2 depends only on $K(\cdot)$. On the other hand, since $\mathbf{P}(n \in \tau) \le Cn^{\alpha-1}$ for a suitable constant $C = C(K(\cdot))$ [cf. (A.11)] we have, for $b_j - a_j = 0$ or 1

$$\sum_{n=(b_j-a_j-1)k\vee 0}^{(b_j-a_j+1)k} \mathbf{P}(n \in \tau) \le C \sum_{n=(b_j-a_j-1)k\vee 0}^{(b_j-a_j+1)k} n^{\alpha-1} \le 2C\alpha^{-1} \frac{k^\alpha}{(b_j-a_j+1)^{1-\alpha}}, \tag{7.37}$$

where the last inequality holds for k larger than a suitable constant. For $b_j - a_j = 2,3,\dots$ we have instead

$$\sum_{n=(b_j-a_j-1)k\vee 0}^{(b_j-a_j+1)k} \mathbf{P}(n \in \tau)$$

$$\le C \sum_{n=(b_j-a_j-1)k}^{(b_j-a_j+1)k} n^{\alpha-1} \le \frac{2Ck^\alpha}{(b_j-a_j-1)^{1-\alpha}} \le \frac{2C3^{1-\alpha}k^\alpha}{(b_j-a_j+1)^{1-\alpha}}. \tag{7.38}$$

We now put (7.37) and (7.38) together to see that

$$\sum_{n=(b_j-a_j-1)k\vee 0}^{(b_j-a_j+1)k} \mathbf{P}(n \in \tau) \le c_3 \frac{k^\alpha}{(b_j-a_j+1)^{1-\alpha}}, \tag{7.39}$$

with $c_3 = c_3(K(\cdot))$.

Now we go back to (7.33): by neglecting the last term which is smaller than one and by using (7.35), (7.36) and (7.39) we obtain

$$\mathbf{P}(E(\mathscr{I})) \leq c_1(k)(c_2c_3)^p \frac{1}{k^\alpha a_1^{1+\alpha}} \prod_{j=1}^{p-1} \frac{1}{(a_{j+1}-b_j)^{1+\alpha}(b_j-a_j+1)^{1-\alpha}}$$

$$\leq C_1 C_2^p \frac{1)}{k^\alpha a_1^{1+\alpha}} \prod_{j=1}^{p-1} \frac{1}{(a_{j+1}-b_j)^{1+\alpha}}, \quad (7.40)$$

with $C_1 = C_1(k) = c_1(k)/k^\alpha$ and $C_2 = c_2c_3$. Therefore the proof of Lemma 7.2 is complete(Fig. 7.1). □

References

1. G. Giacomin, H. Lacoin, F.L. Toninelli, Marginal relevance of disorder for pinning models. Commun. Pure Appl. Math. **63**, 233–265 (2010)
2. G. Giacomin, H. Lacoin, F.L. Toninelli, Disorder relevance at marginality and critical point shift. Ann. Inst. H. Poincaré (B) Probab. Stat. **47**, 148–175 (2011)
3. F.L. Toninelli, Coarse graining, fractional moments and the critical slope of random copolymers. Electron. J. Probab. **14**, 531–547 (2009)

Chapter 8
Path Properties

Abstract We present a few selected results that show that localization, respectively delocalization, in the free energy sense does correspond to localized, respectively delocalized, path behavior. We provide also an overview of the literature on path behavior for disorder pinning models, with a particular attention to the main theme of these notes, that is disorder (ir)relevance, and therefore with a particular eye to trying to quantify the differences in path behavior between pure and disordered systems. While certain questions have found satisfactory answers, important (and intriguing) issues remain open.

8.1 Overview

This chapter is less comprehensive than the other ones, in the sense that we aim at giving the state of the art in the field, but we detail just a few selected results.

We are going to split the presentation according to whether $\mathrm{F}(\beta,h) > 0$ or not and, in both cases, we start by stating and proving the selected results and then we move to the review-like part of the text. We are essentially going to assume $\beta > 0$ because for $\beta = 0$ one can get much more detailed results. However the case $\beta = 0$, cf. Theorem 2.5, plays a central role in this section because it is the reference case: the precision of the results available for the $\beta = 0$ case is of course difficult to achieve, but the important point is that it is precisely in trying to match these results that substantial differences between $\beta = 0$ and $\beta > 0$ appear and this is of course the most intriguing, still only (very) partially understood, aspect of this chapter.

The starting elementary observation is of course that for $h < h_c(\beta)$ we have $\partial_h \mathrm{F}(\beta,h) = 0$ and this is also the case for $h_c(\beta)$ if the free energy $\mathrm{F}(\beta,\cdot)$ is C^1 (and this is the case for example when we can apply the smoothing inequality, cf. Chap. 5). As already pointed out before, in itself this is a rather satisfactory statement because it tells us that the path behavior differs for (β,h) in the localized

G. Giacomin, *Disorder and Critical Phenomena Through Basic Probability Models*, Lecture Notes in Mathematics 2025, DOI 10.1007/978-3-642-21156-0_8,
© Springer-Verlag Berlin Heidelberg 2011

region and in (the interior of) the delocalized one. But the limits of such a statement are rather evident (possibly more in the delocalized case than in the localized one):

1. In the localized regime one observes a positive contact density and, hence, something like a recurrent renewal (possibly even with exponential inter-arrival law). This vague statement contains (at least) two issues: the existence of the infinite volume limit and its closeness to a renewal with exponential inter-arrival law (is the exponential rate still read off the free energy itself?).
2. In the delocalized regime the zero contact density is a rather poor statement in itself if we think of the case $\alpha < 1$ or the case in which the free renewal is transient. And it gets even poorer if one thinks again to the fact that if $\beta = 0$ the contact set is a finite set if $h < h_c = h_c(0)$. So the basic issue is to go beyond the $o(N)$ result.

We deal with the localized regime before, but even before we take a brief excursus through measure concentration inequalities.

8.2 A Quick Look at Concentration Inequalities

The theory of concentration of measures has grown to be a central tool in the statistical mechanics of disordered systems, e.g. [6]. The inequality that we are going to use applies to Lipschitz functions: more precisely it applies to $G : \mathbb{R}^N \to \mathbb{R}$ which is such that for every x and $y \in \mathbb{R}^N$

$$|G(x) - G(y)| \leq C_{\mathrm{Lip}} |x - y| , \tag{8.1}$$

for C_{Lip} a positive constant (and $|x-y|$ is the Euclidean norm). We always work under the Hypothesis 3.1 but we have to add further conditions:

1. Either ω_1 is a bounded variable.
2. Or the law of ω_1 satisfies a log-Sobolev inequality, that is if for every non-negative C^1 function $f : \mathbb{R} \to \mathbb{R}$ such that $\mathbb{E}[f(\omega_1)] = 1$ there exists $\rho \in (0, \infty)$ such that

$$\mathbb{E}\left[f(\omega_1) \log f(\omega_1)\right] \leq \frac{1}{2\rho} \mathbb{E}\left[\frac{(f'(\omega_1))^2}{f(\omega_1)}\right] . \tag{8.2}$$

Such a condition is satisfied in particular if ω_1 is Gaussian (and much more, see e.g. [3, Chap. 6] and [9, Appendix A.3]).

The concentration inequality we are going to use says that for ω as above there exist two positive constants c_1 and c_2 such that for every $G(\cdot)$ convex that satisfies (8.1) we have that $\mathbb{E}|G(\omega_1, \omega_2, \ldots, \omega_N)| < \infty$ and for every $t \geq 0$

$$\mathbb{P}\left(|G(\omega_1, \omega_2, \ldots, \omega_N) - \mathbb{E}G(\omega_1, \omega_2, \ldots, \omega_N)| \geq t\right) \leq c_1 \exp\left(-c_2 t^2 / C_{\mathrm{Lip}}^2\right) . \tag{8.3}$$

For a proof see for example [15, Corollary 4.10 and Proposition 1.8] for the case of bounded variables and [15, Chap. 5] for the log-Sobolev case. There are many other important references on concentration, we signal here [20] and [23].

A by now standard application of concentration inequality to statistical mechanics is the quantitative self-averaging estimate that relies on the observation that $\log Z_{N,\omega}$ is a (convex) Lipschitz function with $C_{\mathrm{Lip}} = \beta\sqrt{N}$: if, for ω and $\omega' \in \mathbb{R}^N$, we set $\omega(s) := s\omega + (1-s)\omega'$

$$\begin{aligned}
\left|\log Z_{N,\omega} - \log Z_{N,\omega'}\right| &= \left|\int_0^1 \frac{\mathrm{d}}{\mathrm{d}s} \log Z_{N,\omega(s)}\, \mathrm{d}s\right| \\
&= \beta \left|\int_0^1 \sum_{n=1}^N (\omega_n - \omega_n')\mathbf{E}_{N,\omega(s)}\left[\delta_n\right] \mathrm{d}s\right| \\
&\le \beta \sqrt{\sum_{n=1}^N (\omega_n - \omega_n')^2} \sqrt{\sum_{n=1}^N \sup_s \mathbf{E}_{N,\omega(s)}\left[\delta_n\right]^2} \le \beta\sqrt{N}|\omega - \omega'|,
\end{aligned} \tag{8.4}$$

where, in the third step, we have used the Cauchy–Schwarz inequality.

Therefore, by (8.3), we have

$$\mathbb{P}\left(\left|\frac{1}{N}\log Z_{N,\omega} - \frac{1}{N}\mathbb{E}\log Z_{N,\omega}\right| \ge \frac{t}{\sqrt{N}}\right) \le c_1 \exp\left(-c_2 t^2/\beta^2\right), \tag{8.5}$$

which is telling in particular that the fluctuations of the random variable $\log Z_{N,\omega}$ are $O(\sqrt{N})$ (this is actually sharp in the localized phase where one can prove a Central Limit Theorem with non-degenerate variance, see [12]). Note also that such an inequality reduces the existence of the quenched free energy to the existence of the quenched averaged free energy, that is to the existence of $\lim_N (1/N)\mathbb{E}\log Z_{N,\omega}$ (proven in the beginning of Chap. 3).

In order to get some interesting results in the delocalized phase we need to go beyond (8.4): we are in fact going to apply the steps in (8.4) to a *restricted partition function*. For $m \in \mathbb{N}$ we introduce the event $E_m := \{|\tau \cap (0,N]| = m\}$ by proceeding like in (8.4) we obtain that for $m \in \{1,2,\ldots,N\}$

$$\begin{aligned}
&\left|\log Z_{N,\omega}(E_m) - \log Z_{N,\omega'}(E_m)\right| \\
&\quad \le \beta \sqrt{\sum_{n=1}^N (\omega_n - \omega_n')^2} \sqrt{\sup_s \sum_{n=1}^N \mathbf{E}_{N,\omega(s)}\left[\delta_n \middle| E_m\right]^2} \le \beta\sqrt{m}|\omega - \omega'|,
\end{aligned} \tag{8.6}$$

where in the last step we have used $\mathbf{E}_{N,\omega(s)}\left[\delta_n \middle| E_m\right] \le 1$ and that $\sum_n \mathbf{E}_{N,\omega(s)}\left[\delta_n \middle| E_m\right]$ is m. Therefore a direct application of (8.3) yields

$$\mathbb{P}\left(\left|\log Z_{N,\omega}(E_m) - \mathbb{E}\log Z_{N,\omega}(E_m)\right| \ge t\right) \le c_1 \exp\left(-c_2 \frac{t^2}{\beta^2 m}\right), \tag{8.7}$$

which is telling us that $\log Z_{N,\omega}(E_m)$ has fluctuations of the order of $\sqrt{m}$ and this will be crucial in Sect. 8.4.

8.3 The Localized Regime

8.3.1 A Basic Observation (and its Consequences)

Here is an argument developed in a slightly different context [19] that translates into mathematics the intuitive idea that if $\mathrm{F}(\beta,h)>0$ then the paths we observe keep close to to zero because those are the paths that contribute to the partition function (rather: if they don't keep close to zero, they do not contribute). This argument makes naturally appear a companion, that we call $\mu(\beta,h)$, to the free energy $\mathrm{F}(\beta,h)$ (a role for this new quantity has been first pointed out in [2]).

In order to develop the argument in the simplest context let us consider a system that extends from $-N$ to N, that is the system with partition function

$$Z_{-N,N,\omega,\beta,h}=Z_{-N,N,\omega}=\mathbf{E}_{-N}\left[\exp\left(\sum_{n=-N+1}^{N}(\beta\omega_n+h)\,\delta_n\right)\delta_N\right], \tag{8.8}$$

where $\mathbf{P}_{-N}$ is the law of the delayed renewal with *delay* $\tau_0=-N$. Our aim is to show that the probability that the origin is in a (large) region that does not contain contact points is small in the localized regime. For this let us introduce the random variable $\mathrm{gap}(\tau):=\max\{l+r:\,l,r\geq 0,\,[-l,r]\cap\tau=\{-l,r\}\}$, so that $\mathrm{gap}(\tau)=0$ if $0\in\tau$ and otherwise $\mathrm{gap}(\tau)>1$. For $g=2,3,\ldots$ by the renewal property of τ we see that we can write

$$\mathbf{P}_{-N,N,\omega}\left(\mathrm{gap}(\tau)\geq g\right)=\sum_{l,r}\frac{Z_{-N,-l,\omega}K(l+r)\exp(\beta\omega_r+h)Z_{r,N,\omega}}{Z_{-N,N,\omega}}, \tag{8.9}$$

where l and r are in $\{1,\ldots,N\}$ and satisfy $l+r\geq g$. But, for every choice of l and r in the range of summation, we have also

$$Z_{-N,N,\omega}\geq Z_{-N,N,\omega}\left(\{-l,r\}\subset\tau\right)=Z_{-N,-l,\omega}Z_{-l,r,\omega}Z_{r,N,\omega}, \tag{8.10}$$

so that

$$\mathbf{P}_{-N,N,\omega}\left(\mathrm{gap}(\tau)\geq g\right)\leq\sum_{l,r}\frac{K(l+r)\exp(\beta\omega_r+h)}{Z_{-l,r,\omega}}, \tag{8.11}$$

so that the term in the sum is really *confined* to $\{-l,\ldots,r\}$ (N is forgotten!). Since

$$\lim_{g\to\infty}\max_{l,r:\,l+r\geq g}\left|\frac{1}{l+r}\log Z_{-l,r,\omega}-\mathrm{F}(\beta,h)\right|=0\quad\mathbb{P}(\mathrm{d}\omega)\text{-a.s.}, \tag{8.12}$$

is an easy corollary to Theorem 3.2, we see, by recalling also that the law of large number implies that if $\omega_1 \in L^1$ then $\lim_{N\to\infty} \sup_{n=1,\dots,N} \omega_n/N = 0$ $\mathbb{P}(\mathrm{d}\omega)$-a.s., that for every $\varepsilon > 0$ and every ω we can find $C_\varepsilon(\omega)$, with $\mathbb{P}(C_\varepsilon(\omega) \in (0,\infty)) = 1$, such that

$$Z_{-l,r,\omega} \exp(-\beta\omega_r - h) \geq \frac{1}{C_\varepsilon(\omega)} \exp\left((l+r)\left(\mathrm{F}(\beta,h) - \varepsilon\right)\right), \tag{8.13}$$

for every l and r in $\mathbb{N}$. Such a statement is of course of interest only if $\mathrm{F}(\beta,h) > 0$, which we assume henceforth, along with $\varepsilon \leq \mathrm{F}(\beta,h)/2$. Going back to (8.11) we therefore see that

$$\begin{aligned} \mathbf{P}_{-N,N,\omega}(\mathrm{gap}(\tau) \geq g) &\leq C_\varepsilon(\omega) \sum_{\substack{l,r:\\ l+r\geq g}} K(l+r) \exp\left(-(l+r)\left(\mathrm{F}(\beta,h) - \varepsilon\right)\right) \\ &\leq C_\varepsilon(\omega) \sum_{j=g}^{\infty} jK(j) \exp\left(-j\left(\mathrm{F}(\beta,h) - \varepsilon\right)\right) \\ &\leq C'_\varepsilon(\omega) \exp\left(-g\left(\mathrm{F}(\beta,h) - \varepsilon\right)\right) \end{aligned} \tag{8.14}$$

where $C'_\varepsilon(\omega)$ is just $C_\varepsilon(\omega)$ times a constant, easily computed, that depends only only on $K(\cdot)$ and on $\mathrm{F}(\beta,h) - \varepsilon (> 0)$.

The bound (8.14) is quite strong and we will discuss it in Sect. 8.5. Here we want to address the following issue: what happens if one considers the *quenched-averaged* measure $\mathbb{E}\mathbf{P}_{-N,N,\omega}$ rather than the quenched measure $\mathbf{P}_{-N,N,\omega}$. Up to now we did not feel the need to make a difference between quenched and quenched-averaged because at the level of the free energy there is no difference. Let us see what happens to the argument we have developed: from (8.11) we have

$$\begin{aligned} \mathbb{E}\mathbf{P}_{-N,N,\omega}(\mathrm{gap}(\tau) \geq g) &\leq \sum_{l,r} K(l+r) \mathbb{E}\left[\frac{\exp(\beta\omega_r + h)}{Z_{-l,r,\omega}}\right] \\ &= \frac{\exp(h)}{\mathrm{M}(-\beta)} \sum_{l,r} K(l+r) \mathbb{E}\left[\frac{1}{Z_{-l,r,\omega}}\right], \end{aligned} \tag{8.15}$$

where we recall that the summation is over $l + r \geq g$. In a sense now the estimate is easier than before, because we are now dealing with a non-random quantity and it would be essentially the same as before if $\log \mathbb{E}\left[1/Z_{-l,r,\omega}\right]$ were close to $-\mathrm{F}(\beta,h)(l+r)$ for $l+r$ large: we will see that it is not so. It is however an elementary exercise [absolutely analogous to the proof of (3.6)] to see that $\{\log \mathbb{E}\left[1/Z_{r,\omega}\right]\}_r$ is sub-additive, so that

$$\lim_{r\to\infty} \frac{1}{r} \log \mathbb{E}\left[\frac{1}{Z_{r,\omega}}\right] = \inf_r \frac{1}{r} \log \mathbb{E}\left[\frac{1}{Z_{r,\omega}}\right] =: -\mu(\beta,h). \tag{8.16}$$

We sum up what we have proven.

Proposition 8.1. *For every $\varepsilon > 0$*

1. *And $\mathbb{P}(\mathrm{d}\omega)$-a.s. there exists $C_\varepsilon(\omega) \in (0,\infty)$ such that*

$$\mathbf{P}_{-N,N,\omega}\left(\mathrm{gap}(\tau) \geq g\right) \leq C_\varepsilon(\omega) \exp\left(-g\left(\mathrm{F}(\beta,h) - \varepsilon\right)\right), \tag{8.17}$$

for every N and every g.

2. *There exists $C_\varepsilon \in (0,\infty)$ such that*

$$\mathbb{E}\mathbf{P}_{-N,N,\omega}\left(\mathrm{gap}(\tau) \geq g\right) \leq C_\varepsilon \exp\left(-g\left(\mu(\beta,h) - \varepsilon\right)\right), \tag{8.18}$$

for every N and every g.

This statement is of course empty if $(\beta,h) \notin \mathscr{L}$: this is clear for the quenched statement, but it is clear also for the quenched averaged case since Jensen inequality directly implies

$$\mu(\beta,h) \leq \mathrm{F}(\beta,h). \tag{8.19}$$

We will see in a moment that, for rather general charge distributions, $\mathrm{F}(\beta,h) > 0$ implies $\mu(\beta,h) > 0$, so that also Proposition 8.1(2) is really a statement for $(\beta,h) \in \mathscr{L}$.

Still, what we have proven up to here is still a weak argument in favor of a substantial role for $\mu(\beta,h)$, but the point is that Proposition 8.1 can be upgraded with moderate effort to a version with lower bounds matching the upper bounds, notably for $(\beta,h) \in \mathscr{L}$ and every $\varepsilon > 0$ one can find $c_\varepsilon > 0$ such that

$$\mathbb{E}\mathbf{P}_{-N,N,\omega}\left(\mathrm{gap}(\tau) \geq g\right) \geq c_\varepsilon \exp\left(-g\left(\mu(\beta,h) + \varepsilon\right)\right), \tag{8.20}$$

for every N and every $g \leq N$ (see [12] and [9, Theorem 7.5]). But in Sect. 8.5 we will come back with an even more stringent reason in favor of the importance of $\mu(\beta,h)$.

8.3.2 On $\mu(\beta,h)$ and $\mathrm{F}(\beta,h)$

We have already seen that $\mu(\beta,h) \leq \mathrm{F}(\beta,h)$. Another quick result is that $\mu(\beta,h) \geq 0$ for every β and h: this is just obtained by using $Z_{N,\omega} \geq \exp(\beta\omega_N + h)K(N)$. The following results are however less immediate.

Proposition 8.2. *The following two statements hold.*

1. *If $\omega_1 \sim \mathcal{N}(0,1)$ (see however Remark 8.3 for generalizations), for $\beta > 0$ and $(\beta,h) \in \mathscr{L}$ we have $\mu(\beta,h) < \mathrm{F}(\beta,h)$.*
2. *If ω is such that the concentration inequality (8.3) holds, then there exists $c > 0$ such that $\mu(\beta,h) \geq c\min\left(\mathrm{F}(\beta,h), \mathrm{F}^2(\beta,h)/\beta^2\right)$, so that $\mathrm{F}(\beta,h) > 0$ implies $\mu(\beta,h) > 0$.*

Proof. For every fixed $N \in \mathbb{N}$ and $\varepsilon > 0$ we call $\widetilde{\mathbb{P}}_N$ the law of the sequence $\omega_1 - \varepsilon/\beta, \ldots, \omega_N - \varepsilon/\beta, \omega_{N+1}, \omega_{N+2}, \ldots$. It is immediate to verify that

$$\frac{1}{N} H\left(\widetilde{\mathbb{P}}_N \middle| \mathbb{P}\right) = \frac{\varepsilon^2}{2\beta^2}, \tag{8.21}$$

and therefore, by Jensen inequality and change of variable, we get

$$\begin{aligned} \frac{1}{N}\log \mathbb{E}\left[\frac{1}{Z_{N,\omega,\beta,h}}\right] &= \frac{1}{N}\log \widetilde{\mathbb{E}}_N\left[\frac{1}{Z_{N,\omega,\beta,h}}\exp\left(\log\frac{\mathrm{d}\mathbb{P}}{\mathrm{d}\widetilde{\mathbb{P}}_N}(\omega)\right)\right] \\ &\geq -\frac{1}{N}\widetilde{\mathbb{E}}_N\left[\log Z_{N,\omega,\beta,h}\right] - \frac{1}{N}H\left(\widetilde{\mathbb{P}}_N\middle|\mathbb{P}\right) \\ &= -\frac{1}{N}\mathbb{E}\left[\log Z_{N,\omega,\beta,h-\varepsilon}\right] - \frac{\varepsilon^2}{2\beta^2}. \end{aligned} \tag{8.22}$$

In the $N \to \infty$ limit we obtain

$$\mu(\beta,h) \leq \mathrm{F}(\beta,h-\varepsilon) + \frac{\varepsilon^2}{2\beta^2}. \tag{8.23}$$

But convexity and the fact that $\mathrm{F}(\beta,h) > 0$ ensure that there exists $c > 0$ such that $\mathrm{F}(\beta,h-\varepsilon) \leq \mathrm{F}(\beta,h) - c\varepsilon$ for ε sufficiently small (choose for example $2c = \lim_{\varepsilon\searrow 0}(\mathrm{F}(\beta,h) - \mathrm{F}(\beta,h-\varepsilon))/\varepsilon$). By coupling this observation and (8.23) we are done with the first statement.

For the second statement we introduce the event

$$\Omega_N := \left\{\omega : \frac{1}{N}\log Z_{N,\omega} \geq \frac{1}{2}\mathrm{F}(\beta,h)\right\}. \tag{8.24}$$

For $(\beta,h) \in \mathscr{L}$ we can find N_0 such that

$$\frac{1}{N}\mathbb{E}\log Z_{N,\omega} \geq \frac{3}{4}\,\mathrm{F}(\beta,h), \tag{8.25}$$

for every $N \geq N_0$. So if we make such an assumption on N the concentration bound (8.5) tells us that

$$\mathbb{P}\left(\Omega_N^{\complement}\right) \leq c_1 \exp\left(-\frac{c_2}{32\beta^2}N\,\mathrm{F}(\beta,h)^2\right), \tag{8.26}$$

that we employ as follows:

$$\begin{aligned} \mathbb{E}\left[\frac{1}{Z_{N,\omega,\beta,h}}\right] &\leq \mathbb{E}\left[\frac{1}{Z_{N,\omega,\beta,h}};\,\Omega_N\right] + \frac{\mathbb{E}\left[\exp(-\beta\omega_N - h);\,\Omega_N^{\complement}\right]}{K(N)} \\ &\leq \exp\left(-\frac{1}{2}N\,\mathrm{F}(\beta,h)\right) + cN^{1+\alpha}\exp\left(-\frac{c_2}{32\beta^2}N\,\mathrm{F}(\beta,h)^2\right), \end{aligned} \tag{8.27}$$

with $c = c_1 \sup_N N^{1+\alpha}/K(N)$. Since $\log(a+b) \le \log 2 + \log\max(a,b)$ for $a,b \ge 0$, the proof is complete. □

Remark 8.3. The Gaussian charge assumption in Theorem 8.2(1) is directly generalized to the cases in which one can exhibit a bound on the relative entropy in (8.21) that goes like ε^a for some $a > 1$. This of course requires that the support of ω_1 is the whole real line. But one can deal also with bounded charges, by proceeding like in [11].

8.4 The Delocalized Regime

Here is the result we want to establish.

Theorem 8.4. *Besides the basic assumption (Hypothesis 3.1) on the law of the charges we require also that the charges have been chosen so that* (8.3) *holds (this includes the case of bounded and Gaussian charges). For every $h < h_c(\beta)$ there exists a constant $C > 0$ such that*

$$\mathbb{E}\mathbf{P}_{N,\omega}\left(|\tau \cap (0,N]| \ge M\right) \le C N^{1+\alpha} \exp(-M/C), \tag{8.28}$$

for every N and $M \in \mathbb{N}$.

The main application of this result is by choosing $M = q \log N$, with $q > C(1+\alpha)$, so that this result shows that, for the *quenched averaged* measure $\mathbb{E}\mathbf{P}_{N,\omega}$, there are $O(\log N)$ pinned sites in the delocalized regime. See Sect. 8.5 for more.

Proof. As we have pointed out in Chap. 3

$$(\beta,h) \in \mathscr{D} \iff \mathbb{E}\left[\log Z_{N,\omega,\beta,h}\right] \le 0 \text{ for every } N \in \mathbb{N}, \tag{8.29}$$

which is just a restatement of Proposition 3.4. Recall now that E_m, introduced just before (8.6), is the event on which there are precisely m contacts and below $Z_{N,\omega}(E)$ is the partition function restricted to renewal trajectories in E. Since if $(\beta,h) \in \mathring{\mathscr{D}}$ then $(\beta,h+\varepsilon) \in \mathscr{D}$ for some $\varepsilon > 0$, we can couple the observation that

$$\log Z_{N,\omega,\beta,h}(E_m) \ge -\frac{1}{2}\varepsilon m \iff \log Z_{N,\omega,\beta,h+\varepsilon}(E_m) \ge \frac{1}{2}\varepsilon m, \tag{8.30}$$

with the fact that $\mathbb{E}\log Z_{N,\omega,\beta,h+\varepsilon}(E_m) \le 0$, which follows from $Z_{N,\omega,\beta,h+\varepsilon}(E_m) \le Z_{N,\omega,\beta,h+\varepsilon}$, to see that

$$\mathbb{P}\left(\log Z_{N,\omega,\beta,h}(E_m) \geq -\frac{1}{2}\varepsilon m\right) = \mathbb{P}\left(\log Z_{N,\omega,\beta,h+\varepsilon}(E_m) \geq \frac{1}{2}\varepsilon m\right)$$
$$\leq \mathbb{P}\left(\log Z_{N,\omega,\beta,h+\varepsilon}(E_m) - \mathbb{E}\log Z_{N,\omega,\beta,h+\varepsilon}(E_m) \geq \frac{1}{2}\varepsilon m\right)$$
$$\leq c_1 \exp\left(-\frac{c_2\varepsilon^2}{4\beta^2} m\right), \quad (8.31)$$

and in the last step we have applied (8.7). Such an estimate directly entails that if we set

$$\Omega_M := \left\{\omega : \text{there exists } m \geq M \text{ such that } \log Z_{N,\omega,\beta,h}(E_m) \geq -\frac{1}{2}\varepsilon m\right\}, \quad (8.32)$$

we have

$$\mathbb{P}(\Omega_M) \leq \sum_{m\geq M} c_1 \exp\left(-\frac{c_2\varepsilon^2}{4\beta^2} m\right) \leq \frac{1}{C_1}\exp(-C_1 M), \quad (8.33)$$

for a suitable choice of $C_1 > 0$ (easily made explicit).

We are now ready to estimate the expectation that appears in (8.28). If $\omega \in \Omega_M^{\complement}$ we have

$$\mathbf{P}_{N,\omega,\beta,h}(|\tau \cap (0,N]| \geq M) = \frac{1}{Z_{N,\omega,\beta,h}} \sum_{m\geq M} Z_{N,\omega,\beta,h}(E_m)$$
$$\leq C_2 \exp(-\beta\omega_N - h) N^{1+\alpha} \sum_{m\geq M} \exp(-\varepsilon m/2) \quad (8.34)$$
$$\leq C_3 \exp(-\beta\omega_N) N^{1+\alpha} \exp(-\varepsilon M/2).$$

Therefore

$$\mathbb{E}\mathbf{P}_{N,\omega,\beta,h}(|\tau \cap (0,N]| \geq M) \leq C_4 N^{1+\alpha} \exp(-\varepsilon M/2) + \frac{1}{C_1}\exp(-C_1 M), \quad (8.35)$$

and we are done. □

8.5 Path Behavior: Overview of What is Known and What is Not

A proper outline of the research on pinning models cannot be given without mentioning the companion class of models called *copolymers and selective solvents* or *copolymer near a selective interface* [5, 19]. And in fact copolymers have already appeared in these notes, but when coming to path properties copolymer and pinning literatures become substantially more entangled. This is due to the fact that it has been in the context of copolymer models, precisely in [5], that the physical approach of looking at the free energy, leaving (temporarily) aside the path properties, has been first taken up by mathematicians. For a *simultaneous* treatment of copolymer and pinning models see [9], notably Chaps. 7 and 8 for what concerns path properties. Here we review the literature with a particular attention to the issues closest to the subject of these notes (pinning, disorder relevance). Let us also point out that it is not clear whether a suitable scaling limit of the disordered pinning model leads to a non-trivial (i.e. disordered) limit model: the (non-trivial!) weak coupling limit [5, 7] of copolymer models becomes trivial for pinning models [18].

8.5.1 On the Localized (and Critical) Regime

Copolymer (localized) path behavior has been studied in [2, 19], before the free energy definition of localization had been given. Properties of copolymer trajectories that are localized in the free energy sense have been first studied in [4] and then, both for copolymers and pinning, in [12]. In particular one can find in [4, 12] statements about the existence and uniqueness of $\mathbf{P}_{\infty,\omega}$, its Gibbs characterization and decay of spatial correlations.

In a sense the results that have been established show that when the free energy is positive a strong form of localization holds: one cannot prove a statement as precise as Proposition 2.9, but, as we have seen in this chapter, in the localized phase large gaps between pinned sites are exponentially penalized and a large localized disordered system roughly looks like a large localized homogeneous system. The reader may therefore have the impression that the difference is simply that in the disordered case one cannot write explicitly the limit model. This is actually not the case and subtle, but important, differences arise when one looks carefully: notably, the new *free energy* $\mu(\beta,h)$, cf. (8.16), pops up naturally as soon as $\beta > 0$. Proposition 8.1 (and the observation that the upper bounds it provides can be matched by analogous lower bounds, see [12] and [9, Chap. 7]) tells us that for large N the measures $\mathbb{E}\mathbf{P}_{N,\omega}$ and $\mathbf{P}_{N,\omega}$ are substantially (or, at least, quantitatively) different. Therefore disorder does play an *active* role on paths. We sum this up here by saying that $1/\mathrm{F}(\beta,h)$ is the correlation length of the quenched model, while $1/\mu(\beta,h)$ is the correlation length of the quenched averaged model [13, 22]: as we have seen, they are different, but are they *substantially* different? Namely, do their behavior approaching criticality differs? The analogous question for Ising models has been taken up in [8]. The appearance of more than one natural correlation length in disordered systems is a very intriguing issue that would lead us far, but we want to point out that in [13] it is shown that the critical exponent of $\mu(\beta,h)$, for $h \searrow h_c(\beta)$, coincides with the one of $\mathrm{F}(\beta,h)$ when disorder is irrelevant, so the two correlation lengths are essentially one! In [13] one can find stronger results on this issue. The open and very interesting question is: can one show that $\mu(\beta,h)$ and $\mathrm{F}(\beta,h)$ have different critical behaviors when disorder is relevant? This of course hits the major open question of understanding the criticality of disordered (pinning) models when disorder is relevant (see [1, 16, 24] for some intriguing conjectures).

We signal also the estimates on the number of pinned sites at criticality given in [14, 21] and the proof, in [14], that at the critical point annealed and critical models are very close to each other in a a.s. pathwise sense in the irrelevant disorder regime.

In [12] and [9, Chap. 7] (but also in [2] for copolymers) one can find a proof (and a precise statement) of the fact that the largest gap between pinned points in a disordered system of size N (gap(τ) of Sect. 8.3 is just the gap containing the origin) is, to leading order, of size $\mu(\beta, h)\log N$, both for $\mathbf{P}_{N,\omega}$ and $\mathbb{E}\mathbf{P}_{N,\omega}$ (this quantity is self averaging!). This is one more signature of the disorder: in the homogeneous case the largest gap goes like $\mathrm{F}(0,h)\log N$ and it is directly tight to the decay of the probability that gap(τ) is large. For more on this see [13].

8.5.2 On the Delocalized Regime

The $O(\log N)$ results in Theorem 8.4 (proven in [10]) can be improved to $O(1)$ results when (β, h) are in the delocalized regime of the annealed model, essentially because one can exploit directly the homogeneous system estimates (see [10] and [9, Chap. 8]). Here we point out that, thanks to Remarks 6.4 and 6.6, one can directly extend these results to the whole regime in which the fractional moment method applies (does this regime coincide with the full delocalized non-critical regime?).

The common characteristics (and drawback) of all the results we have mentioned about the delocalized regime (recall however [14] for the critical case) is that they are all results about the measure $\mathbb{E}\mathbf{P}_{N,\omega}$, and not about the quenched model itself (see [10] for some considerations on the fact that certain statements that one would guess at first for the delocalized behavior of $\mathbf{P}_{N,\omega}$ cannot hold). But very recently a proof of the fact that Theorem 8.4 can be upgraded to a $\mathbb{P}(\mathrm{d}\omega)$-a.s. statement on $\mathbf{P}_{N,\omega}$ has been found [17].

References

1. A. Aharony, A.B. Harris, Absence of self-averaging and universal fluctuations in random systems near critical points. Phys. Rev. Lett. **77**, 3700–3703 (1996)
2. S. Albeverio, X.Y. Zhou, Free energy and some sample path properties of a random walk with random potential. J. Stat. Phys. **83**, 573–622 (1996)
3. C. Ané, S. Blachère, D. Chafaï, P. Fougères, I. Gentil, F. Malrieu, C. Roberto, G. Scheffer, *Sur les inégalités de Sobolev logarithmiques*. Panoramas et Synthèses **10**, Société Mathématique de France, Paris, 2000
4. M. Biskup, F. den Hollander, A heteropolymer near a linear interface, Ann. Appl. Probab. **9**, 668–687 (1999)
5. E. Bolthausen, F. den Hollander, Localization transition for a polymer near an interface. Ann. Probab. **25**, 1334–1366 (1997)

6. A. Bovier, *Statistical Mechanics of Disordered Systems. A Mathematical Perspective*. Cambridge Series in Statistical and Probabilistic Mathematics (Cambridge University Press, Cambridge, 2006)
7. F. Caravenna, G. Giacomin, The weak coupling limit of disordered copolymer models. Ann. Probab. **38**, 2322–2378 (2010)
8. J.T. Chayes, L. Chayes, D.S. Fisher, T. Spencer, Correlation length bounds for disordered Ising ferromagnets. Commun. Math. Phys. **120**, 501–523 (1989)
9. G. Giacomin, *Random Polymer Models* (Imperial College Press, London, 2007)
10. G. Giacomin, F.L. Toninelli, Estimates on path delocalization for copolymers at selective interfaces. Probab. Theory Relat. Fields **133**, 464–482 (2005)
11. G. Giacomin, F.L. Toninelli, Smoothing effect of quenched disorder on polymer depinning transitions. Commun. Math. Phys. **266**, 1–16 (2006)
12. G. Giacomin, F.L. Toninelli, The localized phase of disordered copolymers with adsorption. ALEA Lat. Am. J. Probab. Math. Stat. **1**, 149–180 (2006)
13. G. Giacomin, F.L. Toninelli, On the irrelevant disorder regime of pinning models. Ann. Probab. **37**, 1841–1873 (2009)
14. H. Lacoin, The martingale approach to disorder irrelevance for pinning models. Electron. Commun. Probab. **15**, 418–427 (2010)
15. M. Ledoux, *The Concentration of Measure Phenomenon*. Mathematical Surveys and Monographs, vol. 89 (American Mathematical Society, Providence, RI, 2001)
16. C. Monthus, T. Garel, Distribution of pseudo-critical temperatures and lack of self-averaging in disordered Poland-Scheraga models with different loop exponents. Eur. Phys. J. B **48**, 393–403 (2005)
17. J.-C. Mourrat, On the delocalized phase of the random pinning model. arXiv:1010.4671
18. N. Pétrélis, Copolymer at selective interfaces and pinning potentials: weak coupling limits. Ann. Inst. H. Poincaré (B) Probab. Stat. **45**, 175–200 (2009)
19. Ya. G. Sinai, *A random walk with a random potential*. Theory Probab. Appl. **38**, 382–385 (1993)
20. M. Talagrand, A new look at independence. Ann. Probab. **24**, 1–34 (1996)
21. F.L. Toninelli, Critical properties and finite-size estimates for the depinning transition of directed random polymers. J. Stat. Phys. **126**, 1025–1044 (2007)
22. F.L. Toninelli, *Localization transition in disordered pinning models. Effect of randomness on the critical properties*, in *Methods of Contemporary Mathematical Statistical Physics*, Lecture Notes in Mathematics, vol. 1970 (2009), pp. 129–176
23. C. Villani, *Topics in Optimal Transportation*. Graduate Studies in Mathematics, vol. 58 (American Mathematical Society, Providence, RI, 2003)
24. S. Wiseman, E. Domany, Finite-size scaling and lack of self-averaging in critical disordered systems. Phys. Rev. Lett. **81**, 22–25 (1998)

Appendix A
Discrete Renewal Theory: Basic (and a Few Less Basic) Facts and Estimates

A.1 A Crash Course on Renewal Theory

A.1.1 Renewal and Markov Chains

We start by working in a general (discrete) framework, that is we choose a discrete probability density $K(\cdot)$ on $\mathbb{N} \cup \{\infty\}$ ($K(\infty) < 1$ to avoid trivialities) and we introduce $\tau := \{\tau_j\}_{j=0,1,\ldots}$ as the sequence of partial sums of an IID sequence of $K(\cdot)$ distributed variables, that we call *inter-arrival variables*. We call τ $K(\cdot)$*-renewal* and we stress that $\tau_0 = 0$ unless explicitly stated. We also freely switch from looking at τ as a sequence of random variables and as a (random) subset of $\mathbb{N} \cup \{0\}$ (*point process*): note that we do not include infinity in this case, because it is always the case that either τ is a finite set (when $\tau_n = \infty$ for some n) or τ contains infinitely many points (but not ∞). The point process notational convention is rather practical and compact: for example $\{$there exists j such that $\tau_j = n\}$ shrinks down to $\{n \in \tau\}$. We say that τ is *persistent* when $|\tau| = \infty$ ($|\tau|$ is the number of points in τ); otherwise we say that it is *terminating*. Of course τ is persistent if and only if $K(\infty) = 0$.

Renewal processes enjoy the renewal property, i.e. if $A \subset \mathscr{P}(\{0,1,\ldots,n\})$ and $B \subset \mathscr{P}(\{n+1, n+2\ldots,\})$ we have

$$\mathbf{P}(\tau \cap [0,n] \in A, n \in \tau, \tau \cap [n+1,\infty) \in B) = \mathbf{P}(\tau \cap [0,n] \in A, n \in \tau)\,\mathbf{P}(\tau + n \in B). \quad \text{(A.1)}$$

There is natural link between renewal processes and Markov chains (see [1, Chap. I] for a quick self-contained review on Markov chains and for all basic notions). In fact, by the strong Markov property the sequence of successive returns of a Markov chain to a fixed (recurrent or transient) state is a renewal process. But this works also in the opposite direction: any renewal process is the return time sequence of a suitable Markov chain. In fact if we define $\mathrm{A}_n := \mathrm{A}_n(\tau) := n - \sup\{\tau_k : \tau_k \le n\}$, then the sequence $\mathrm{A} := \{\mathrm{A}_n\}_{n=0,1,\ldots}$ is a Markov chain called

G. Giacomin, *Disorder and Critical Phenomena Through Basic Probability Models*, Lecture Notes in Mathematics 2025, DOI 10.1007/978-3-642-21156-0,
© Springer-Verlag Berlin Heidelberg 2011

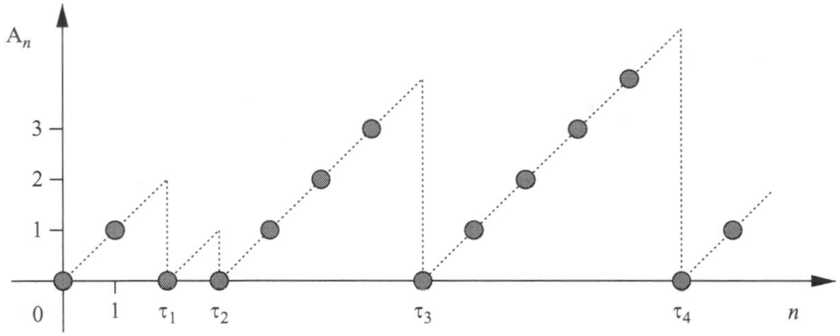

Fig. A.1 The A process (large gray dots) associated to the renewal τ. In this case $\tau_1 = 2$, $\tau_2 = 3$, $\tau_3 = 7$, $\tau_4 = 12$ and $\tau_5 > 13$

backward recurrence time. Note that $\mathrm{A}_n \in \mathbb{N} \cup \{0\}$ is the time elapsed since the last renewal when looking from n, see Fig. A.1. The probability transition from $\mathrm{A}_n = i$ to $\mathrm{A}_{n+1} = j$ is non-zero only if $j = i+1$ or $j = 0$ and the probability that the process moves up is $(\overline{K}(i+1)+K(\infty))/(\overline{K}(i)+K(\infty))$, independently of $\mathrm{A}_0, \mathrm{A}_1, \ldots, \mathrm{A}_{n-1}$ (we have used $\overline{K}(i) := \sum_{n>i} K(n)$ and in this sum $K(\infty)$ is not included).

We say that $K(\cdot)$ has period $p \in \mathbb{N}$ if $\{n : K(n) > 0\}$ is contained in $\{pn : n \in \mathbb{N}\}$ and if p is the largest number with this property. If $p = 1$ we say that $K(\cdot)$ is aperiodic. The aperiodicity of $K(\cdot)$ implies the aperiodicity of the Markov chain A. Note that when $K(\infty) = 0$, the state space of A is $\{0, 1, \ldots, \sup\{n\colon K(n) > 0\}\} \setminus \{\infty\}$ and that A is irreducible (that is all states communicate in a finite number of steps, with positive probability).

For ease of notation set $m_K := \sum_{n \in \mathbb{N} \cup \{\infty\}} nK(n) \in [1, \infty]$. Of course $m_K = \infty$ may arise also when $K(\infty) = 0$: in this case A is a recurrent Markov chain, but it is immediate to see that it is a null recurrent chain, since τ_1 coincides with $\inf\{n > 0 : \mathrm{A}_n = 0\}$. On the other hand, A is clearly positive recurrent if $m_K < \infty$. We will therefore say that τ is positive (respectively null) persistent if A is.

A.1.2 The Renewal Theorem

We state now the Theorem.

Theorem A.1. *If* $K(\cdot)$ *is aperiodic then* $\lim_{n\to\infty} \mathbf{P}(n \in \tau) = 1/m_K$, *with* $1/\infty = 0$.

We direct the reader to [1, Chap. I, Theorem 2.2] for a proof, which is based on the fundamental formula (direct consequence of the renewal property), often called *renewal equation* (and $n \mapsto \mathbf{P}(n \in \tau)$ is the *renewal function*), that says

$$\mathbf{P}(n \in \tau) = \mathbf{1}_{n=0} + \sum_{k=0}^{n} \mathbf{P}(k \in \tau) K(n-k). \tag{A.2}$$

One way to extract information from (A.2) is to pass to Laplace transform: for $s > 0$

$$\sum_{n=0}^{\infty} \exp(-sn)\mathbf{P}(n \in \tau) = \left(1 - \sum_{n=1}^{\infty} \exp(-sn)K(n)\right)^{-1}. \tag{A.3}$$

It helps the intuition to note that Theorem A.1 can be proven also by looking at the ergodic properties of the backward recurrence time process A [1, Sect. VII.2]. In particular one sees that the condition $m_K < \infty$ is precisely the condition for A to be positive recurrent. This implies directly the existence of a unique invariant probability measure which we can write explicitly:

$$p_{\mathrm{A}}(n) = \frac{1}{m_K} \sum_{j \geq n+1} K(j), \quad n = 0, 1, \ldots. \tag{A.4}$$

By the Ergodic Theorem for irreducible aperiodic Markov chains one also has that

$$\lim_{N \to \infty} \mathbf{P}(\mathrm{A}_N = n) = p_{\mathrm{A}}(n). \tag{A.5}$$

If $\mathrm{A} = \{\mathrm{A}_n\}_n$ is redefined so that A_0 is distributed according to $p_{\mathrm{A}}(\cdot)$, so that A is stationary, then $\{n : \mathrm{A}_n = 0\}$ is a renewal process translated by a random quantity (independent of the renewal). In other terms, τ_0 is no longer degenerate, but it has its own distribution (on $\mathbb{N} \cup \{0\}$). We call such a process a *delayed renewal* and when τ_0 has distribution $p_{\mathrm{A}}(\cdot)$ [cf. (A.4)] we call *stationary renewal* such a delayed renewal (note that also a delayed renewal enjoys the renewal property).

We will often need information on the number of renewal points up to a certain time n, that is on $|\tau \cap (0,n]|$. The fact that

$$\lim_{n \to \infty} \frac{1}{n} |\tau \cap (0,n]| = \frac{1}{m_K}, \tag{A.6}$$

both in the almost sure and in the L^1 sense (as a matter of fact, in L^p for any $p \geq 1$), is a direct consequence of Kolmogorov law of large numbers.

A.1.3 Beyond the Renewal Theorem

Theorem A.1 does apply when $m_K = \infty$, but in this case it calls for refinements. We state here results in this direction in the framework in which we typically work: that is, we assume (2.30).

First of all, the transient case is covered by the following result.

Theorem A.2. *If $K(\infty) > 0$ then*

$$\mathbf{P}(N \in \tau) \stackrel{N \to \infty}{\sim} \frac{K(N)}{(K(\infty))^2}. \tag{A.7}$$

A proof can be found in [7, A.5.2].

For the null recurrent case we have of course $\alpha \in (0,1]$. The following results can be found in [2, Theorem 8.7.3 and Theorem 8.7.5].

Theorem A.3. *For $\alpha \in (0,1)$ and $K(\infty) = 0$ we have that*

$$\mathbf{E}[|\tau \cap (0,N]|] = \sum_{n=1}^{N} \mathbf{P}(n \in \tau) \stackrel{N\to\infty}{\sim} \frac{\sin(\pi\alpha)}{c_K \pi} N^{\alpha}. \tag{A.8}$$

For $\alpha = 1$ and $K(\infty) = 0$ we have that $\sum_{n=1}^{N} \mathbf{P}(n \in \tau) \sim N/(c_K \log N)$.

Theorem A.3 gives *integral* bounds and it can be obtained by standard Tauberian arguments [2] applied to (A.3). Getting *local* estimates is substantially harder: sharp local estimates however have been obtained and this is the content of the next theorem.

Theorem A.4. *For $\alpha \in (0,1)$ and $K(\infty) = 0$ we have*

$$\mathbf{P}(N \in \tau) \stackrel{N\to\infty}{\sim} \frac{\alpha \sin(\pi\alpha)}{c_K \pi} \frac{1}{N^{1-\alpha}}. \tag{A.9}$$

For $\alpha = 1$ instead

$$\mathbf{P}(N \in \tau) \stackrel{N\to\infty}{\sim} \frac{1}{c_K \log N}. \tag{A.10}$$

Formula (A.9) is due to Doney [4, Theorem B], that completed a partial result of [6]. The second result instead, that is (A.10), can be found in [2, Theorem 8.7.5].

Of course, from (A.9) one directly extracts the existence for $\alpha \in (0,1)$ of $C = C(K(\cdot)) > 0$ such that for every $N \in \mathbb{N}$

$$\frac{1}{C} \le N^{1-\alpha} \mathbf{P}(N \in \tau) \le C. \tag{A.11}$$

A.1.4 Convergence of Renewal and Point Processes

A renewal or, more generally, a discrete point process can be seen as a random variable $\tau : \Omega \to \mathscr{P}(\mathbb{N} \cup \{0\}) = 2^{\mathbb{N}\cup\{0\}}$, with$(\Omega, \mathscr{F}, \mathbf{P})$ a generic probability space and $\mathscr{P}(\mathbb{N} \cup \{0\})$ is equipped with the σ-algebra $\mathscr{G} := \sigma(\cup_{n\in\mathbb{N}} \mathscr{G}_n))$, where $\{\mathscr{G}_n\}_{n=0,1,\ldots}$ is the natural filtration of the process $\mathrm{A} = \mathrm{A}(\tau)$ defined in Sect. A.1. It is convenient to introduce a metric space for which $\mathscr{G}$ is the Borel σ-algebra. For this we introduce the semi-metric d_n on $\mathscr{P}(\mathbb{N} \cup \{0\})$ by setting $d_n(B_1, B_2) = \mathbf{1}_{B_1 \cap [0,n] \neq B_2 \cap [0,n]}$. Then one directly verifies that $\mathscr{G}$ is the Borel σ-algebra for the metric $d(B_1, B_2) := \sum_n 2^{-n} d_n(B_1, B_2)$. We introduce the notation

$$\mathscr{P}_0 := (\mathscr{P}(\mathbb{N} \cup \{0\}), d), \tag{A.12}$$

for the metric space. In general we do not deal only with renewal processes: the typical case we consider is the one of a sequence of measures $\{\mathbf{P}_N\}_N$ on $\mathscr{P}_0$ converging to a limit (that happens to be a renewal process). In any case the convergence in law of $\{\mathbf{P}_N\}_N$ to a limit $\mathbf{P}_\infty$ means simply if $B \in \mathscr{G}_n$ for some n then $\lim_N \mathbf{P}_N(B) = \mathbf{P}_\infty(B)$.

A.2 Some Pinning Oriented Renewal Issues

A.2.1 On Boundary Effects

Next is a bound on boundary effects: we just state the inequality that we use, but also the reciprocal bound can be proven (with a different constant) along the same line.

Lemma A.5. *Assume* (2.30)*. For every $K(\cdot)$-renewal τ there exists $C_{bc} > 0$ such that*

$$\mathbf{E}\left[F_n(\tau)\,\middle|\,2n \in \tau\right] \le C_{bc}\,\mathbf{E}\left[F_n(\tau)\right], \tag{A.13}$$

for every $n \in \mathbb{N}$ and every $F_n : \mathscr{P}(\mathbb{N}) \longrightarrow [0,\infty)$ ($\mathscr{P}(\mathbb{N})$ is the set of subsets of $\mathbb{N}$) which is measurable with respect to the σ-algebra generated by $\{\{A \in \mathscr{P}(\mathbb{N}) : j \in A\} : j = 1,2,\ldots,n\}$ (that is $F_n(A) = F_n(A \cap (0,n])$ for every $A \subset \mathbb{N}$).

Proof. The various constants c_1, c_2,... appearing below depend only on $K(\cdot)$. Set $X_n = X_n(\tau) := \max\{j = 0,1,\ldots,n : j \in \tau\}$. By the measurability properties of $F_n(\cdot)$ we see that $F_n(\tau) = F_n(\tau \cap (0,j])$ if $X_n(\tau) = j$ (read $(0,0]$ as $\emptyset$) and therefore

$$\mathbf{E}\left[F_n(\tau)\,\middle|\,2n \in \tau\right] = \sum_{j=0}^{n} \mathbf{E}\left[F_n(\tau \cap (0,j])\,\middle|\,X_n = j\right] \mathbf{P}\left(X_n = j\,\middle|\,2n \in \tau\right), \tag{A.14}$$

and therefore it is sufficient to show that

$$\mathbf{P}\left(X_n = j\,\middle|\,2n \in \tau\right) \le C_{bc}\,\mathbf{P}\left(X_n = j\right). \tag{A.15}$$

For this we write

$$\begin{aligned}\mathbf{P}\left(X_n = j,\, 2n \in \tau\right) &= \mathbf{P}\left(j \in \tau\right) \sum_{m=n+1}^{2n} K(m-j)\mathbf{P}\left(2n - m \in \tau\right) \\ &= \mathbf{P}\left(j \in \tau\right)\left(\sum_{m=n+1}^{\lfloor 3n/2 \rfloor} \ldots + \sum_{m=\lfloor 3n/2 \rfloor + 1}^{2n} \ldots\right) =: T_1 + T_2.\end{aligned} \tag{A.16}$$

For what concerns T_1 we use the fact that $2n-m \geq n/2$ so that $\mathbf{P}(2n-m \in \tau) \leq c_1 \mathbf{P}(2n \in \tau)$ and

$$T_1 \leq c_1 \mathbf{P}(2n \in \tau)\mathbf{P}(j \in \tau) \sum_{m=n+1}^{\lfloor 3n/2 \rfloor} K(m-j). \tag{A.17}$$

For T_2 we use that $\mathbf{P}(j \in \tau) \leq c_2(j+1)^{\alpha-1}$ for every j and that $m-j \geq \lfloor n/2 \rfloor$, which implies $K(m-j) \leq c_3/n^{1+\alpha}$, so that

$$T_2 \leq c_2 c_3 \mathbf{P}(j \in \tau) n^{-1-\alpha} \sum_{m=\lfloor 3n/2 \rfloor+1}^{2n} (2n-m+1)^{\alpha-1} \leq c_4 \mathbf{P}(j \in \tau) n^{-1}, \tag{A.18}$$

for every $n \in \mathbb{N}$. Since we have $\mathbf{P}(2n \in \tau) \geq c_5 n^{\alpha-1}$ and $\sum_{m=\lfloor 3n/2 \rfloor+1}^{\infty} K(m-j) \geq c_6 n^{-\alpha}$ (for $j \leq n$), from (A.18) we get that

$$T_2 \leq \frac{c_4}{c_5 c_6} \mathbf{P}(2n \in \tau)\mathbf{P}(j \in \tau) \sum_{m=\lfloor 3n/2 \rfloor+1}^{\infty} K(m-j). \tag{A.19}$$

By putting (A.17) and (A.19) together we have that

$$\begin{aligned} &\mathbf{P}(X_n = j, 2n \in \tau) \\ &\quad \leq c_7 \mathbf{P}(2n \in \tau)\mathbf{P}(j \in \tau) \sum_{m=n+1}^{\infty} K(m-j) = c_7 \mathbf{P}(2n \in \tau)\mathbf{P}(X_n = j), \end{aligned} \tag{A.20}$$

which completes the proof with $C_{bc} = c_7$. □

A.2.2 Two Scaling Results for Renewal Processes

The first result is applied in Chap. 6 when $\alpha \in (1/2, 1)$, but it plays a central role also for the case $\alpha = 1/2$, since it is used in Proposition A.7 below, that, in turn, is used in Chap. 6.

Proposition A.6. *For every $K(\cdot)$-renewal τ with $K(\cdot)$ as in* (2.30) *and $\alpha \in (0,1)$ we have*

$$\mathscr{L}\text{-}\lim_{n\to\infty} \frac{1}{n^\alpha} |\tau \cap (0,n]| = \frac{Y_\alpha}{c_K}, \tag{A.21}$$

where Y_α is a random variable that depends only on α with the property $\mathbf{P}(Y_\alpha > 0) = 1$. In particular $Y_{1/2} = |Z|/\sqrt{2\pi}$, with $Z \sim \mathcal{N}(0,1)$.

Proof. This is treated for example in [5]. The point is simply that $|\tau \cap (0,n]| < m$ is equivalent to $\tau_m > n$ and τ_m is a sum of m IID variables and the question is therefore an issue of domain of stability of stable laws. In [5, XI.5, p. 373] it is proven that for every $x > 0$

$$\lim_{n\to\infty} \mathbf{P}\left(\overline{K}(n)\,|\tau\cap(0,n]| \geq \frac{2-\alpha}{\alpha}\frac{1}{x^\alpha}\right) = G_\alpha(x), \tag{A.22}$$

where $G_\alpha(\cdot)$ is the distribution function of the one sided stable distribution satisfying $\lim_{x\to\infty} x^\alpha(1-G_\alpha(x)) = (2-\alpha)/\alpha$, characterized by the Laplace transform

$$\int_0^\infty \exp(-\lambda x)\mathrm{d}G_\alpha(x) = \exp(-c_\alpha\lambda^\alpha) \quad \text{with } c_\alpha := \frac{(2-\alpha)\Gamma(1-\alpha)}{\alpha}, \tag{A.23}$$

for $\lambda > 0$. Such stable laws are treated in detail for example in [5, XIII.6, Theorem 1], where (A.23) is proven along with the fact that $\lim_{x\searrow 0} G_\alpha(x) = 0$, so that $(0,\infty)$ is of full measure under this distribution (for completeness: in this limit $G_\alpha(x) = o(\exp(-cx^{-\alpha}))$, with $c = \alpha/((2-\alpha)\Gamma(1-\alpha))$). By a change of variable and by using $\overline{K}(n) \sim (\mathtt{c}_K/\alpha)n^{-\alpha}$ in (A.22) we see that

$$\lim_{n\to\infty} \mathbf{P}\left(\frac{|\tau\cap(0,n]|}{n^\alpha} \geq y\right) = G_\alpha\left(\left(\frac{2-\alpha}{\mathtt{c}_K}\right)^{\frac{1}{\alpha}} y^{-\frac{1}{\alpha}}\right) =: 1 - F_\alpha(y), \tag{A.24}$$

and the asymptotic properties of $G_\alpha(x)$ mentioned just above directly yield that $F_\alpha(y)$ tends to 0 as $y \searrow 0$ (more precisely: $F_\alpha(y) \sim \mathtt{c}_K y/\alpha$) and that $\lim_{y\to\infty} F_\alpha(y) = 1$ (more precisely, $1 - F_\alpha(y) = o(\exp(-cy))$ for a $c > 0$). These facts suffice to conclude the converge in law that we claim in the statement and that the limit variable Y_α is a.s. positive. A number of further properties of Y_α can be derived by exploiting the properties of the stable distribution $G_\alpha(\cdot)$, for example that $G_\alpha(t) = \int_0^t g_\alpha(s)\mathrm{d}s$ for a suitable probability density $g_\alpha(\cdot)$ (this follows immediately from the fact that the characteristic function $\psi_\alpha(t) = \int_0^\infty \exp(itx)\mathrm{d}G_\alpha(x) = \exp(-c_\alpha t^\alpha(\cos(\pi\alpha/2) + i\sin(\pi\alpha/2))$ for $t \geq 0$, so that $|\psi_\alpha(t)| = \exp(-c_\alpha|t|^\alpha(\cos(\pi\alpha/2))$ and therefore $\int_{\mathbb{R}} |\psi(t)|\mathrm{d}t < \infty$). However, *it seems impossible to express stable densities in a closed form* [5, p. 581], with the notable exception of $\alpha = 1/2$ for which we can use that if the random variable X has density $f_X(x) = x^{-3/2}(2\sqrt{\pi})^{-1}\exp(-1/(4x))\mathbf{1}_{x>0}$ we have

$$\mathbf{E}\left[\exp(-\lambda X)\right] = \exp\left(-\sqrt{\lambda}\right), \quad \text{for } \lambda > 0. \tag{A.25}$$

A straightforward, but rather painful, constant tracking exercise [via (A.23) and (A.24)] leads to $Y_{1/2} = |Z|/\sqrt{2\pi}$. □

Proposition A.7. *For every $K(\cdot)$-renewal τ with $K(\cdot)$ as in (2.30) and $\alpha = 1/2$ we have*

$$\mathscr{L}\text{-}\lim_{n\to\infty} \frac{1}{\sqrt{n}\log n} \sum_{1\leq i<j\leq n} \frac{\delta_i\delta_j}{\sqrt{j-i}} = \frac{|Z|}{(2\pi)^{3/2}\,\mathtt{c}_K^2}, \tag{A.26}$$

where $Z \sim \mathcal{N}(0,1)$.

Proof. We introduce the notation

$$Y_n^{(i)} := \sum_{j=i+1}^{n} \frac{\delta_j}{\sqrt{j-i}}, \tag{A.27}$$

that allows writing

$$\frac{1}{\sqrt{n}\log n} \sum_{1\le i<j\le n} \frac{\delta_i \delta_j}{\sqrt{j-i}} = \frac{1}{\sqrt{n}\log n} \sum_{i=1}^{n-1} \delta_i Y_n^{(i)} =: X_n\,. \tag{A.28}$$

Note that, by the renewal property of τ, $Y_n^{(i)}$ (under $\mathbf{P}(\cdot\,|\,\delta_i = 1)$) is distributed like $Y_{n-i} := Y_{n-i}^{(0)}$ (under $\mathbf{P}$). The first step in the proof is observing that, by (A.11), we have

$$\mathbf{E}\left[\frac{1}{\sqrt{n}\log n} \sum_{i=(1-\varepsilon)n}^{n-1} \delta_i Y_n^{(i)}\right] = \frac{1}{\log n\sqrt{n}} \sum_{i=(1-\varepsilon)n}^{n-1} \sum_{j=i+1}^{n} \frac{\mathbf{P}(i\in\tau)\mathbf{P}(j-i\in\tau)}{\sqrt{j-i}} = O(\varepsilon), \tag{A.29}$$

uniformly in n: we have introduced the short-cut convention (that we will keep throughout this proof) that summing from $(1-\varepsilon)n$ means summing from $\lfloor(1-\varepsilon)n\rfloor + 1$ and, just below, summing up to $(1-\varepsilon)n$ means up to $\lfloor(1-\varepsilon)n\rfloor$. What (A.29) is telling us is that we can focus on studying $X_{n,\varepsilon}$, defined as X_n, but stopping the sum over i at $(1-\varepsilon)n$:

$$X_{n,\varepsilon} := \frac{1}{\sqrt{n}\log n} \sum_{i=1}^{(1-\varepsilon)n} \delta_i Y_n^{(i)}\,. \tag{A.30}$$

At this point we use that

$$\lim_{n\to\infty} \frac{Y_n}{\log n} = \frac{1}{2\pi c_K} =: \widehat{c}_K, \tag{A.31}$$

in $L^2(\mathbf{P})$ (and hence in $L^1(\mathbf{P})$). We postpone the proof of (A.31) and observe that, since the normalization is the logarithm of n, it implies that for every $\varepsilon \in (0,1)$

$$\lim_{n\to\infty} \sup_{q\in[\varepsilon,1]} \mathbf{E}\left[\left|\frac{1}{\log n} \sum_{j=1}^{\lfloor qn\rfloor} \frac{\delta_j}{\sqrt{j}} - \widehat{c}_K\right|\right] = 0. \tag{A.32}$$

Let us write

$$R_n := X_{n,\varepsilon} - \frac{\widehat{c}_K}{\sqrt{n}} \sum_{i=1}^{(1-\varepsilon)n} \delta_i \tag{A.33}$$

and note that $n^{-1/2}\sum_{i=1}^{(1-\varepsilon)n}\delta_i$ converges in law toward $\sqrt{(1-\varepsilon)/(2\pi c_K^2)}\,|Z|$ by Proposition A.6. It suffices therefore to show that for every $\varepsilon \in (0,1)$ we have $\lim_{n\to\infty}\mathbf{E}[|R_n|] = 0$. And in fact

$$\mathbf{E}[|R_n|] \le \frac{1}{\sqrt{n}}\sum_{i=1}^{(1-\varepsilon)n}\mathbf{E}[\delta_i]\mathbf{E}\left[\left|\frac{Y_n^{(i)}}{\log n} - \widehat{c}_K\right| \,\middle|\, \delta_i = 1\right]$$
$$= \frac{1}{\sqrt{n}}\sum_{i=1}^{(1-\varepsilon)n}\mathbf{E}[\delta_i]\mathbf{E}\left[\left|\frac{Y_{n-i}}{\log n} - \widehat{c}_K\right|\right] \xrightarrow{n\to\infty} 0, \quad \text{(A.34)}$$

where in the last step we have used (A.32) and (A.11).

We are therefore left with proving (A.31). This result has been established in [3, Theorem 6] in the case in which τ is given by the successive returns to zero of a centered, aperiodic and irreducible random walk on $\mathbb{Z}$ with bounded increment variance. Note that, by well established local limit theorems, for such a class of random walks we have (A.9). In [3] it is proven more, namely that (A.31) holds also almost surely and this is extracted from the estimate $\mathrm{var}_{\mathbf{P}}(Y_n) = O(\log n)$. What we are going to do is simply to re-obtain such a bound, by repeating the steps in [3] and using (A.9)–(A.11), for the general renewal processes that we consider (and one can verify that almost sure convergence comes as a bonus, but we will not use it).

The proof goes as follows: by (A.9) one directly sees that $\lim_{n\to\infty}\mathbf{E}[Y_n/\log n] = \widehat{c}_K$, therefore we are done if we show that $\lim_{n\to\infty}\mathrm{var}_{\mathbf{P}}(Y_n/\log n) = 0$. So we start by observing that

$$\mathrm{var}_{\mathbf{P}}(Y_n) = \sum_{i,j}\frac{\mathbf{E}[\delta_i\delta_j] - \mathbf{E}[\delta_i]\mathbf{E}[\delta_j]}{\sqrt{ij}} = 2\sum_{i=1}^{n-1}\sum_{j=i+1}^{n}\frac{\mathbf{E}[\delta_i\delta_j] - \mathbf{E}[\delta_i]\mathbf{E}[\delta_j]}{\sqrt{ij}} + O(1), \quad \text{(A.35)}$$

by (A.11). Now we compute

$$\begin{aligned}\sum_{i=1}^{n-1}\sum_{j=i+1}^{n}\frac{\mathbf{E}[\delta_i\delta_j] - \mathbf{E}[\delta_i]\mathbf{E}[\delta_j]}{\sqrt{ij}} &= \sum_{i=1}^{n-1}\frac{\mathbf{E}[\delta_i]}{\sqrt{i}}\left[\sum_{j=1}^{n-i}\frac{\mathbf{E}[\delta_j]}{\sqrt{j+i}} - \sum_{j=i+1}^{n}\frac{\mathbf{E}[\delta_j]}{\sqrt{j}}\right]\\
&\le \sum_{i=1}^{n-1}\frac{\mathbf{E}[\delta_i]}{\sqrt{i}}\left[\sum_{j=1}^{n-i}\frac{\mathbf{E}[\delta_j]}{\sqrt{j+i}} - \sum_{j=i+1}^{n}\frac{\mathbf{E}[\delta_j]}{\sqrt{j+i}}\right]\\
&\le \sum_{i=1}^{n-1}\frac{\mathbf{E}[\delta_i]}{\sqrt{i}}\sum_{j=1}^{i}\frac{\mathbf{E}[\delta_j]}{\sqrt{j+i}} \le \sum_{i=1}^{n-1}\frac{\mathbf{E}[\delta_i]}{i}\sum_{j=1}^{i}\mathbf{E}[\delta_j]\\
&\le C^2\sum_{i=1}^{n-1}\frac{1}{i^{3/2}}\sum_{j=1}^{i}\frac{1}{j^{1/2}} = O(\log n),\end{aligned} \quad \text{(A.36)}$$

where, in the last line, we have used (A.11). In view of (A.35), we have obtained $\mathrm{var}_{\mathbf{P}}(Y_n) = O(\log n)$ so that the proof (A.31) is complete and, with it, the proof of Lemma A.7. □

A.2.3 On the Derivatives of the Free Energy Near Criticality

We have seen that, assuming (2.30) and $\sum_n K(n) = 1$, for $\alpha \in (0,1)$ we have

$$\mathrm{F}(h) \overset{h \searrow 0}{\sim} c h^{1/\alpha} =: \mathrm{F}_{\mathrm{cr}}(h), \tag{A.37}$$

where $c = (\alpha/(\mathrm{c}_K \Gamma(1-\alpha)))^{1/\alpha} > 0$ (cf. Theorem 2.10). The subscript *cr* is used to indicate that the function captures the leading critical behavior. Recall that $\mathrm{F}(\cdot)$ is real analytic except at the origin. Here we prove that:

Proposition A.8. *For $\alpha \in (0,1)$ and $1/\alpha \notin \mathbb{N}$ we have that for every $j \in \mathbb{N}$*

$$\left(\frac{\mathrm{d}}{\mathrm{d}h}\right)^j \mathrm{F}(h) \overset{h \searrow 0}{\sim} \left(\frac{\mathrm{d}}{\mathrm{d}h}\right)^j \mathrm{F}_{cr}(h) = c h^{-j+1/\alpha} \prod_{i=1}^{j} \left(\frac{1}{\alpha} - i + 1\right). \tag{A.38}$$

If $1/\alpha \in \mathbb{N}$ then (A.38) *holds for $j \le 1/\alpha$.*

This result largely suffices for our purposes, but let us point out that generalizing the statement to $j > 1/\alpha$ when $1/\alpha \in \mathbb{N}$ requires more on $K(\cdot)$ than (2.30).

Proof. Let us start by setting up some notation:

$$\Psi(x) \overset{x \ge 0}{=} 1 - \sum_{n=1}^{\infty} K(n) \exp(-nx) \overset{x \searrow 0}{\sim} \mathrm{c}_K \frac{\Gamma(1-\alpha)}{\alpha} x^{\alpha} =: \Psi_{\mathrm{cr}}(x). \tag{A.39}$$

Let us recall that the relation defining $\mathrm{F}(h)$ for $h > 0$ is

$$\Psi(\mathrm{F}(h)) = 1 - \exp(-h) \overset{h \searrow 0}{\sim} h. \tag{A.40}$$

This formula has the important companion:

$$\Psi_{\mathrm{cr}}(\mathrm{F}_{\mathrm{cr}}(h)) = h. \tag{A.41}$$

In the sequel we use the notation $f^{(j)}(h) := (\mathrm{d}/\mathrm{d}h)^j f(h)$ and we point out that a standard Riemann sum approximation yields for $j \in \mathbb{N}$:

$$\begin{aligned} \Psi^{(j)}(x) &= (-1)^{j+1} \sum_n n^j K(n) \exp(-nx) \overset{x \searrow 0}{\sim} (-1)^{j+1} \Gamma(j-\alpha) \mathrm{c}_K x^{\alpha - j} \\ &= \left(\prod_{i=1}^{j-1} (\alpha - i)\right) \Gamma(1-\alpha) \mathrm{c}_K x^{\alpha - j} = \Psi_{\mathrm{cr}}^{(j)}(x), \end{aligned} \tag{A.42}$$

with the convention $\prod_{i=1}^{0} (\alpha - i) = 1$.

Let us now compute the asymptotic behavior of $\mathrm{F}'(h)$ and of $\mathrm{F}''(h)$: this will serve the double purpose of getting acquainted with the general case and of serving to verify the first step in the induction argument for the general case. First of all from the relation (A.40) defining $\mathrm{F}(h)$ we have

$$\frac{\mathrm{d}}{\mathrm{d}h}\Psi(\mathrm{F}(h)) = \Psi'(\mathrm{F}(h))\mathrm{F}'(h) = \exp(-h) \overset{h\searrow 0}{\sim} 1, \tag{A.43}$$

so that

$$\mathrm{F}'(h) \overset{h\searrow 0}{\sim} \frac{1}{\Psi'(\mathrm{F}(h))} \sim \frac{1}{\Psi'_{\mathrm{cr}}(\mathrm{F}_{\mathrm{cr}}(h))} = \mathrm{F}'_{\mathrm{cr}}(h), \tag{A.44}$$

where the last equality follows by using (A.37) and (A.42) or (more easily!) by taking the derivative of (A.41). For what concerns $\mathrm{F}''(h)$ we compute and use once again the relation (A.40) to get

$$\left(\frac{\mathrm{d}}{\mathrm{d}h}\right)^2 \Psi(\mathrm{F}(h)) = \Psi''(\mathrm{F}(h))\left(\mathrm{F}'(h)\right)^2 + \Psi'(\mathrm{F}(h))\mathrm{F}''(h) \overset{h\searrow 0}{\sim} -1. \tag{A.45}$$

By (A.42) and (A.44) we see that

$$\Psi''(\mathrm{F}(h))\left(\mathrm{F}'(h)\right)^2 \overset{h\searrow 0}{\sim} \frac{c}{h}, \tag{A.46}$$

with $c \neq 0$, so that

$$\Psi''(\mathrm{F}(h))\left(\mathrm{F}'(h)\right)^2 \overset{h\searrow 0}{\sim} -\Psi'(\mathrm{F}(h))\mathrm{F}''(h), \tag{A.47}$$

from which we extract the asymptotic relation of $\mathrm{F}''(h) \sim \mathrm{F}''_{\mathrm{cr}}(h)$ by using again (A.42) and (A.44), that is by using the asymptotic behavior of F, F', Ψ' and Ψ''. Note however that, once again, there is a much cheaper way to go from (A.47) to $\mathrm{F}''(h) \sim \mathrm{F}''_{\mathrm{cr}}(h)$: by taking two derivatives of (A.41) one obtains (A.47) with the subscripts *cr* added (six subscripts in total) and with $\sim$ replaced by $=$, but we already know that we can add the subscripts in (A.47) without altering the validity of the statement to all the functions except a priori F'', and this implies $\mathrm{F}''(h) \sim \mathrm{F}''_{\mathrm{cr}}(h)$.

We have therefore established the claim for $j = 1$ and 2, but explicit expressions for the jth derivative of the composition of two functions are rather involved for arbitrary j. To get the result we want we will go around this point (much in the spirit of the alternative approach used twice above) by observing that

$$\begin{aligned}\left(\frac{\mathrm{d}}{\mathrm{d}h}\right)^j \Psi(\mathrm{F}(h)) &= \Psi^{(1)}(\mathrm{F}(h))\mathrm{F}^{(j)}(h) \\ &\quad + P_j\left(\mathrm{F}^{(1)}(h), \ldots, \mathrm{F}^{(j-1)}(h), \Psi^{(1)}(\mathrm{F}(h)), \ldots, \Psi^{(j)}(\mathrm{F}(h))\right),\end{aligned} \tag{A.48}$$

where P_j is a polynomial: actually $P_1(x)$ is just zero [cf. (A.43)] and $P_2(x,y,z) = zx^2$ [cf. (A.45)], and that (A.48) has the companion

$$\left(\frac{\mathrm{d}}{\mathrm{d}h}\right)^j \Psi_{\mathrm{cr}}(\mathrm{F}_{\mathrm{cr}}(h)) = \Psi_{\mathrm{cr}}^{(1)}(\mathrm{F}_{\mathrm{cr}}(h))\mathrm{F}_{\mathrm{cr}}^{(j)}(h)$$
$$+P_j\left(\mathrm{F}_{\mathrm{cr}}^{(1)}(h), \ldots, \mathrm{F}_{\mathrm{cr}}^{(j-1)}(h), \Psi_{\mathrm{cr}}^{(1)}(\mathrm{F}_{\mathrm{cr}}(h)), \ldots, \Psi_{\mathrm{cr}}^{(j)}(\mathrm{F}(h))\right). \tag{A.49}$$

But the expression in (A.49) is equal to one if $j = 1$ and it is equal to zero if $j \geq 2$, while the analogous expression without the *cr* subscripts takes the value one, or minus one, as $h \searrow 0$. Therefore in this limit

$$\Psi^{(1)}(\mathrm{F}(h)) - \Psi_{\mathrm{cr}}^{(1)}(\mathrm{F}_{\mathrm{cr}}(h)) = O(1). \tag{A.50}$$

For simplicity below we write $P_j(\mathrm{F}^{(1)}, \ldots)$ and $P_j(\mathrm{F}_{\mathrm{cr}}^{(1)}, \ldots)$ for the more cumbersome complete expressions.

We start now the induction procedure that consists in obtaining that $\mathrm{F}^{(j)}(h) \sim \mathrm{F}_{\mathrm{cr}}^{(j)}(h)$ for all $j < \widehat{n}$, with $\widehat{n} = 1 + 1/\alpha$ if $1/\alpha \in \mathbb{N}$ and $\widehat{n} = \infty$ otherwise, knowing that $\mathrm{F}^{(k)}(h) \sim \mathrm{F}_{\mathrm{cr}}^{(k)}(h)$ for $k = 0, 1, \ldots, j-1$. Note that this follows if we can show that

$$P_j(\mathrm{F}^{(1)}(h), \ldots) \overset{h \searrow 0}{\sim} P_j(\mathrm{F}_{\mathrm{cr}}^{(1)}(h), \ldots), \tag{A.51}$$

and that these expressions diverge in this limit. This is because by using (A.50) and the fact that the asymptotic behaviors of the functions in the right-hand sides of (A.48) and (A.49) are all known, except for $\mathrm{F}^{(j)}(h)$ that is then necessarily the same as the one of $\mathrm{F}_{\mathrm{cr}}^{(j)}(h)$ and we are done.

Let us therefore establish (A.51) and the fact that the quantities in it diverge. That they diverge can be established by observing that for $2 \leq j < \widehat{n}$

$$P_j(\mathrm{F}_{\mathrm{cr}}^{(1)}(h), \ldots) = -\Psi_{\mathrm{cr}}^{(1)}(\mathrm{F}_{\mathrm{cr}}(h))\mathrm{F}_{\mathrm{cr}}^{(j)}(h) = c\,h^{1-j}, \tag{A.52}$$

where $c \neq 0$ is an explicit constant (note that if $1/\alpha \in \mathbb{N}$ then $\mathrm{F}_{\mathrm{cr}}^{(j)}(h) = 0$ for $j > 1/\alpha$). For what concerns (A.51) we start by observing that the leading behavior of each of the term constituting $P_j(\mathrm{F}^{(1)}, \ldots)$ (P_j is of course a sum of monomials) is the same of the corresponding term in $P_j(\mathrm{F}_{\mathrm{cr}}^{(1)}(h), \ldots)$. All these monomial terms are of the same order (in a strong sense: any ratio converge to a non-zero value): this can be either verified on the expression with or without the subscript *cr* (in a rather illogical way we verify it for the quantities without subscript: the prof is slightly more involved, since we need to use the induction assumption, but formulas are lighter without subscripts). In fact in taking the derivative that builds $P_j(\dots)$ we repeat two types of operations:

1. Taking a derivative of $\Psi^k(\mathrm{F}(h))$ for $k \le j$ and for this we have

$$\frac{\mathrm{d}}{\mathrm{d}h}\Psi^k(\mathrm{F}(h)) = \Psi^{k+1}(\mathrm{F}(h))\mathrm{F}'(h) \overset{h\searrow 0}{\sim} -(j+1-\alpha)\Psi^k(\mathrm{F}(h))\frac{\mathrm{F}'(h)}{\mathrm{F}(h)} \overset{h\searrow 0}{\sim} -\left(\frac{j-1+\alpha}{\alpha}\right)\frac{\Psi^k(\mathrm{F}(h))}{h}, \tag{A.53}$$

   that is such a derivative makes the term more singular (by a factor $1/h$).
2. Taking derivatives of $\mathrm{F}^{(k)}(h)$ for $k = 1, \dots, j-2$ (by definition of P_j, the derivative of $\mathrm{F}^{(j-1)}(h)$ does not enter P_j): but the asymptotic behaviors of $\mathrm{F}^{(k)}(h)$ and $\mathrm{F}^{(k+1)}(h)$ are in the induction assumption and one directly verifies that $h\mathrm{F}^{(k+1)}(h)/\mathrm{F}^{(k)}(h)$ tends to a non-zero constant (recall that $k < j < \widehat{n}$). Once again, such a derivative asymptotically just introduces a multiplicative $1/h$ factor (times a non-zero constant).

If we now recall that in the starting step ($j = 2$) of the induction we had just two terms, cf. (A.45), and each of order $1/h$ [cf. (A.45)] we see that $P_j(\mathrm{F}^{(1)}(h), \dots)$ (and $P_j(\mathrm{F}_{\mathrm{cr}}^{(1)}(h), \dots)$) is a sum of terms of order h^{1-j}, so that (A.52) is telling us that the asymptotic behavior of $P_j(\mathrm{F}_{\mathrm{cr}}^{(1)}(h), \dots)$ is of the same order of each of the monomial terms that constitute it (and it is not the result of the cancellation between the leading orders of larger terms). Therefore (A.51) holds for $j < \widehat{n}$ and the proof is complete. □

References

1. S. Asmussen, *Applied Probability and Queues*, 2nd edn. (Springer, New York, 2003)
2. N.H. Bingham, C.M. Goldie, J.L. Teugels, *Regular Variation* (Cambridge University Press, Cambridge, 1987)
3. K.L. Chung, P. Erdös, Probability limit theorems assuming only the first moment I. Mem. Am. Math. Soc. **6**, 1–19 (1951), paper 3
4. R.A. Doney, One-sided local large deviation and renewal theorems in the case of infinite mean. Probab. Theory Relat. Fields **107**, 451–465 (1997)
5. W. Feller, *An Introduction to Probability Theory and Its Applications*, vol. II, 2nd edn. (Wiley, New York, 1971)
6. A. Garsia, J. Lamperti, A discrete renewal theorem with infinite mean. Comment. Math. Helv. **37**, 221–234 (1963)
7. G. Giacomin, *Random Polymer Models* (Imperial College Press, London, 2007)

Index

annealed bound, 32
annealed model, 33, 42, 63

backward recurrence time, 114
boundary condition, 6, 35, 117

change of measure estimates, 63, 64, 74, 88
charge, 29
charge distribution, 30
coarse graining, 41, 64, 68, 76, 91, 98
concentration properties, 36, 39, 102, 106, 108
contact density, 10
copolymer, 60, 109
correlation length, 19, 38, 56, 64, 76, 110
critical behavior, 18, 41, 42, 56, 111, 122
critical curve, 33
critical exponent, 43, 56
critical point, 19, 32, 34, 42, 56, 63

defect, 21
delocalization, 10, 32, 101
diluted Ising model, 41, 55, 58
directed polymer, 21
disordered pinning model, 29

entropy-energy competition, 35

fractional moment estimates, 64, 65, 87
free energy, 8, 11, 15, 19, 30, 32, 122
free pinning model, 6

Harris criterion, 41, 55
homogeneous pinning model, 14

inter-arrival law, 7, 14, 113
interface, 21
irrelevant disorder, 41, 46, 58, 110
Ising model, 21, 41, 55

localization, 10, 32, 101

marginal disorder, 42, 58
Markov chain, 113

null persistent renewal, 114

order of phase transition, 19

partition function, 6, 15, 29
path properties, 11, 17, 101
persistent renewal, 7, 45, 113
phase transition, 11, 18, 19, 56
pinning model, 5, 14, 29
Poland-Scheraga model, 24
polymer pinning, 21
positive persistent renewal, 7, 114
pure model, 41

quenched disorder, 29

random walk pinning model, 5
rare stretch strategy, 51
relative entropy, 54, 108
relevant disorder, 42, 58, 111

G. Giacomin, *Disorder and Critical Phenomena Through Basic Probability Models*,
Lecture Notes in Mathematics 2025, DOI 10.1007/978-3-642-21156-0,
© Springer-Verlag Berlin Heidelberg 2011

renewal function, 7
renewal process, 7, 113
renewal property, 113
renewal theorem, 114
renormalization, 41
replica, 44

self-averaging property, 32
smoothing inequality, 51, 60
super-additive property, 30, 34, 35, 48, 54

terminating renewal, 7, 44, 113

List of participants

40th Probability Summer School, Saint-Flour, France

July 4–17, 2010

Lecturers

Franco FLANDOLI	Università di Pisa, Italy
Giambattista GIACOMIN	Université Paris Diderot, France
Takashi KUMAGAI	Kyoto University, Japan

Participants

Sergio ALMADA	Georgia Inst. Technology, Atlanta, USA
Marek ARENDARCZYK	Univ. Wroclaw, Poland
David BARBATO	Univ. Padova, Italy
David BELIUS	ETH Zurich, Switzerland
Pierre BERTIN	Univ. Paris 6 et 7, F
Luigi Amedeo BIANCHI	Scuola Normale Superiore, Pisa, Italy
Thomas BOUILLOC	Univ. Nice Sophia Antipolis, F
Omar BOUKHADRA	U. Provence, F & U. Constantine, Algeria
Charles-Edouard BREHIER	ENS Cachan, Antenne Bretagne, Rennes, F
Elisabetta CANDELLERO	Graz Univ. Technology, Austria
Francesco CARAVENNA	Univ. Padova, Italy
Reda CHHAIBI	Univ. Pierre et Marie Curie, Paris, F
Mirko D'OVIDIO	Sapienza Univ. Roma, Italy
Latifa DEBBI	Univ. Setif, Algeria
François DELARUE	Univ. Nice Sophia Antipolis, F
Francisco DELGADO	Univ. Barcelona, Spain
Aurélien DEYA	Univ. Henri Poincaré, Nancy, F
Hacène DJELLOUT	Univ. Blaise Pascal, Clermont-Ferrand, F
Aurélien EBERHARDT	Univ. Strasbourg, F
François EZANNO	Univ. de Provence, Marseille, F
Ennio FEDRIZZI	Univ. Paris Diderot, F
Matthieu FELSINGER	Univ. Bielefeld, Germany
Robert FITZNER	Eindhoven Univ. Technology, NL
Elena ISSOGLIO	Friedrich Schiller Univ., Jena, Germany

G. Giacomin, *Disorder and Critical Phenomena Through Basic Probability Models*,
Lecture Notes in Mathematics 2025, DOI 10.1007/978-3-642-21156-0,

© Springer-Verlag Berlin Heidelberg 2011

Shuai JING	Univ. Bretagne Occidentale, Brest, F
Yasmina KHELOUFI	Univ. Setif, Algeria
Konrad KOLESKO	Univ. Wroclaw, Poland
Noemi KURT	TU Berlin, Germany
Kazumasa KUWADA	Ochanomizu Univ., Tokyo, Japan
Mateusz KWASNICKI	Wroclaw Univ. Technology, Poland
Clément LAURENT	Univ. de Provence, Marseille, F
Qian LIN	Univ. Bretagne Occidentale, Brest, F
Arnaud LIONNET	Univ. Oxford, UK
J.-A. LOPEZ-MIMBELA	CIMAT, Guanajuato, Mexico
Eric LUÇON	Univ. Pierre et Marie Curie, Paris, F
Camille MALE	ENS Lyon, F
Mario MAURELLI	Univ. Pisa, Italy
Francesco MORANDIN	Univ. Parma, Italy
Jean-Christophe MOURRAT	Univ. de Provence, Marseille, F
Mikhail NEKLYUDOV	Univ. York, UK
Eyal NEUMANN	Technion Inst. Technology, Haifa, Israel
Harald OBERHAUSER	TU Berlin, Germany
Jean PICARD	Univ. Blaise Pascal, Clermont-Ferrand, F
K. PIETRUSKA-PALUBA	Univ. Warsaw, Poland
Marco ROMITO	Univ. Firenze, Italy
Erwan SAINT LOUBERT BIÉ	Univ. Blaise Pascal, Clermont-Ferrand, F
Martin SAUER	TU Darmstadt, Germany
Georg SCHOECHTEL	TU Darmstadt, Germany
Laurent SERLET	Univ. Blaise Pascal, Clermont-Ferrand, F
Yuhao SHEN	Univ. Pierre et Marie Curie, Paris, F
Mykhaylo SHKOLNIKOV	Stanford Univ., USA
Damien SIMON	Univ. Pierre et Marie Curie, Paris, F
Julien SOHIER	Univ. Paris Diderot, F
Philippe SOSOE	Princeton Univ., USA
Andrzej STOS	Univ. Blaise Pascal, Clermont-Ferrand, F
E. TODOROVA KOLKOVSKA	CIMAT, Guanajuato, Mexico
Dario VINCENZI	Univ. Nice Sophia Antipolis, F
Jing WANG	Purdue Univ., West Lafayette, USA
Frédérique WATBLED	Univ. Bretagne-Sud, Vannes, F
Hendrik WEBER	Univ. Warwick, UK
Lihu XU	Eindhoven Univ. Technology, NL
Ramon XULVI-BRUNET	Harvard Univ., Cambridge, MA, USA
Danyu YANG	Oxford Univ., UK
Lorenzo ZAMBOTTI	Univ. Pierre et Marie Curie, Paris, F

LECTURE NOTES IN MATHEMATICS

Springer

Edited by J.-M. Morel, B. Teissier; P.K. Maini

Editorial Policy (for the publication of monographs)

1. Lecture Notes aim to report new developments in all areas of mathematics and their applications - quickly, informally and at a high level. Mathematical texts analysing new developments in modelling and numerical simulation are welcome.
   Monograph manuscripts should be reasonably self-contained and rounded off. Thus they may, and often will, present not only results of the author but also related work by other people. They may be based on specialised lecture courses. Furthermore, the manuscripts should provide sufficient motivation, examples and applications. This clearly distinguishes Lecture Notes from journal articles or technical reports which normally are very concise. Articles intended for a journal but too long to be accepted by most journals, usually do not have this "lecture notes" character. For similar reasons it is unusual for doctoral theses to be accepted for the Lecture Notes series, though habilitation theses may be appropriate.

2. Manuscripts should be submitted either online at www.editorialmanager.com/lnm to Springer's mathematics editorial in Heidelberg, or to one of the series editors. In general, manuscripts will be sent out to 2 external referees for evaluation. If a decision cannot yet be reached on the basis of the first 2 reports, further referees may be contacted: The author will be informed of this. A final decision to publish can be made only on the basis of the complete manuscript, however a refereeing process leading to a preliminary decision can be based on a pre-final or incomplete manuscript. The strict minimum amount of material that will be considered should include a detailed outline describing the planned contents of each chapter, a bibliography and several sample chapters.
   Authors should be aware that incomplete or insufficiently close to final manuscripts almost always result in longer refereeing times and nevertheless unclear referees' recommendations, making further refereeing of a final draft necessary.
   Authors should also be aware that parallel submission of their manuscript to another publisher while under consideration for LNM will in general lead to immediate rejection.

3. Manuscripts should in general be submitted in English. Final manuscripts should contain at least 100 pages of mathematical text and should always include

   – a table of contents;
   – an informative introduction, with adequate motivation and perhaps some historical remarks: it should be accessible to a reader not intimately familiar with the topic treated;
   – a subject index: as a rule this is genuinely helpful for the reader.

   For evaluation purposes, manuscripts may be submitted in print or electronic form (print form is still preferred by most referees), in the latter case preferably as pdf- or zipped psfiles. Lecture Notes volumes are, as a rule, printed digitally from the authors' files. To ensure best results, authors are asked to use the LaTeX2e style files available from Springer's web-server at:

   ftp://ftp.springer.de/pub/tex/latex/svmonot1/ (for monographs) and
   ftp://ftp.springer.de/pub/tex/latex/svmultt1/ (for summer schools/tutorials).

Additional technical instructions, if necessary, are available on request from lnm@springer.com.

4. Careful preparation of the manuscripts will help keep production time short besides ensuring satisfactory appearance of the finished book in print and online. After acceptance of the manuscript authors will be asked to prepare the final LaTeX source files and also the corresponding dvi-, pdf- or zipped ps-file. The LaTeX source files are essential for producing the full-text online version of the book (see http://www.springerlink.com/openurl.asp?genre=journal&issn=0075-8434 for the existing online volumes of LNM). The actual production of a Lecture Notes volume takes approximately 12 weeks.

5. Authors receive a total of 50 free copies of their volume, but no royalties. They are entitled to a discount of 33.3 % on the price of Springer books purchased for their personal use, if ordering directly from Springer.

6. Commitment to publish is made by letter of intent rather than by signing a formal contract. Springer-Verlag secures the copyright for each volume. Authors are free to reuse material contained in their LNM volumes in later publications: a brief written (or e-mail) request for formal permission is sufficient.

Addresses:

Professor J.-M. Morel, CMLA,
École Normale Supérieure de Cachan,
61 Avenue du Président Wilson, 94235 Cachan Cedex, France
E-mail: morel@cmla.ens-cachan.fr

Professor B. Teissier, Institut Mathématique de Jussieu,
UMR 7586 du CNRS, Équipe "Géométrie et Dynamique",
175 rue du Chevaleret
75013 Paris, France
E-mail: teissier@math.jussieu.fr

For the "Mathematical Biosciences Subseries" of LNM:

Professor P. K. Maini, Center for Mathematical Biology,
Mathematical Institute, 24-29 St Giles,
Oxford OX1 3LP, UK
E-mail : maini@maths.ox.ac.uk

Springer, Mathematics Editorial, Tiergartenstr. 17,
69121 Heidelberg, Germany,
Tel.: +49 (6221) 487-8259

Fax: +49 (6221) 4876-8259
E-mail: lnm@springer.com